I0065411

GiveIQ

PURPOSE AND PHILANTHROPY
in the
AGE OF AI

JOHN JUDGE

Dedicated to MPJ and CMPJ

COPYRIGHT © 2026 JOHN JUDGE

TABLE OF CONTENTS

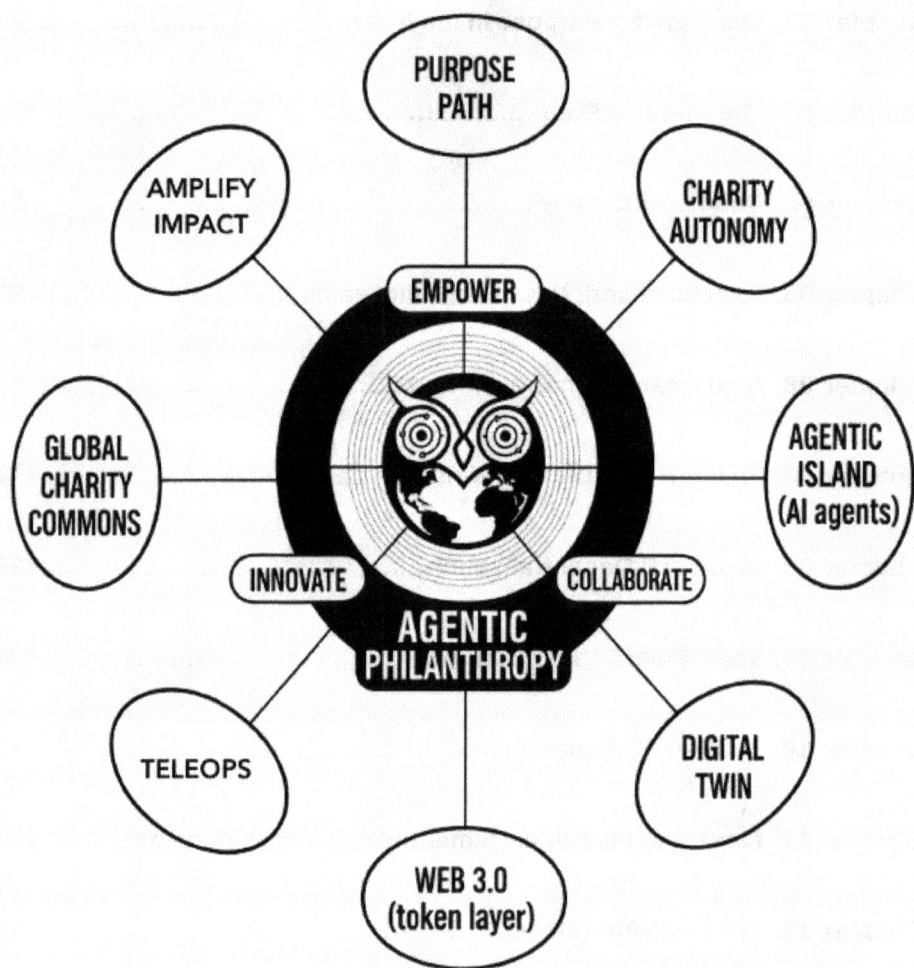

PREFACE

Our Launch Pad: The Age of Purpose Awaits

The central premise of this book is simple: **Human Agency multiplied by Agentic Philanthropy = Greater Societal and Cause Impact.** That equation isn't just a formula for doing more good. It's a survival strategy for the era we're entering.

The tools we have today are merely the beginning. This is the age of early artificial intelligence, the initial forms of Web 3.0, the first indications of quantum computing, and an expansive frontier of tools we have yet to envision. Biological breakthroughs will extend human life. Neuromorphic systems will process information like brains. Humanoid robots will walk among us. The pace of change will not slow down.

In the midst of such hyper-change, what keeps us grounded? What prevents us from being swept away—overwhelmed by information, paralyzed by possibility, missing in the metaverse, or hollowed out by automation?

Purpose does. Philanthropy does. The deliberate choice to orient your life toward contribution rather than consumption—that becomes your anchor when everything else is shifting.

This is why the GiveIQ equation matters now more than ever. Human agency— your irreplaceable capacity to choose, to care, to act with intention—remains the constant. Agentic philanthropy—AI tools that amplify your giving, extend your reach, and multiply your impact—becomes the accelerant. Together, they don't just produce greater cause impact. They produce a more grounded you.

The question is no longer whether we can make a significant difference. The question is whether we will.

I believe the wave of AI, and the tools it inspires, will force each of us to choose: to contribute, or to consume. Now is the time to set personal and societal guardrails. Not to resist technology, but to align it with what it means to live with purpose, to be fully human, and to give.

Stigmergy: How We Once Found Our Way

There's a word for the invisible coordination that makes cairns work: stigmergy.

The term comes from the Greek *stigma* (mark, sign) and *ergon* (work, action). It describes how sophisticated collective outcomes can arise without central leadership, without meetings, without explicit coordination. The environment itself becomes the medium of collaboration.

Watch an ant colony. No ant knows the master plan. No ant receives instructions from headquarters. Yet together, millions of ants build elaborate structures, find efficient paths to food, and respond to threats with remarkable coordination. Each ant simply responds to the traces others have left—pheromone trails that say *this way* or *food here* or *danger*. She follows strong signals, reinforces them with her own traces, and collective intelligence emerges from accumulated small acts.

But stigmergy isn't only for insects. Wikipedia is stigmergy: editors responding to what previous editors left behind. Open-source software is stigmergy: coders building on each other's contributions. The desire path worn across a campus lawn is stigmergic—each walker responding to and reinforcing the traces left by others, a new route emerging that no planner designed.

The cairn is perhaps the purest human example. Complex navigation across generations, through terrain no single person fully knows, arising without leadership or coordination. The system self-organizes through accumulated traces in the environment.

This is how civil society once worked. This is how we learned purpose. This is how philanthropy perpetuated itself across generations.

And this is what we are losing.

The Broken Cairns

I have seen this dearth of community and civil society organizations firsthand.

For the past several years, as the Scout Executive for Greater Boston, I've worked to start new Scouting programs in suburbs and cities across the region. The contrast when I first worked for Scouting – some thirty-five years ago - is stark and sobering.

A generation ago, when you wanted to launch a Scout troop, you knew where to go. There was an active American Legion Post on Main Street. A Knights of Columbus Hall. A Ladies Auxiliary that met on Tuesday nights. A Parent Teacher Organization with real institutional memory and a bench of willing volunteers. There were more people with more time to give—and, crucially, there were *visible examples* of what giving looked like.

A lot of that was thanks to the incredible sacrifice and ethos of the men and women from the World War II generation. They came home from the war and served their communities like no generation before or since. They ran the pancake breakfasts and the scholarship funds. They showed up for the funerals of veterans no one else remembered. They mentored, they modeled, they marked the trail.

We learned from them. They set their stones on the civic cairns. Purpose and philanthropy were the lessons.

Today, those cairns are crumbling. The American Legion has lost nearly half its members since 1992. Mainline church denominations have declined by a third or more. The Elks, the Moose, the Odd Fellows—organizations that once anchored community life in every small town—are shadows of themselves, their remaining members aging, their lodges closing. Bowling leagues, as Robert Putnam famously documented, have collapsed. Union halls sit empty. Even youth sports leagues increasingly operate as transactional service providers rather than communities of mutual obligation.

Each closure, each lapsed membership, each empty chair at the monthly meeting is a stone removed from the cairn.

And here's what that means: fewer traces for those coming after.

A young person today who feels a stirring toward service, toward purpose, toward something larger than herself—where does she look? The pheromone trails have faded. The cairns that once marked the path are scattered, overgrown, invisible. She might have the same instincts her great-grandparents had, the same longing for meaning and connection, but the stigmergic infrastructure that would have guided them is gone.

Some never find the path at all. They sense something is missing but can't name it, can't locate it. They don't reject purpose—they simply never encounter the traces that would lead them toward it.

Others start out but get lost. Without visible cairns, without elders whose example they can follow, without the accumulated wisdom of those who walked before, they wander. They try things. They bounce between causes and apps and moments of enthusiasm that don't accumulate into anything lasting.

They place stones, but in random places. Nothing accumulates. No path emerges.

The trails have faded. But they are still there, ready to be rebuilt and used again.

A Once-in-a-Millennium Opportunity

Artificial intelligence is advancing at an astonishing pace, and most of us will soon be surprised—maybe overwhelmed—by what happens next.

Let me start by saying I am an optimist, sometimes to a fault. I believe this time of AI and technological transformation is a once-in-a-millennium opportunity. Not because the technology is magical, but because it arrives at precisely the moment we need it most.

The old stigmergic infrastructure of civil society has collapsed. The cairns are broken. The trails are fading. A generation is searching for purpose and finding no clear path.

And now, for the first time in human history, we have tools that can make invisible traces visible again. Tools that can connect isolated givers into communities of shared purpose. Tools that can illuminate which approaches are working, where resources are flowing, where the gaps remain. Tools that can help anyone—not just the wealthy, not just the experts—give with intelligence, intention, and impact.

While AI will undoubtedly bring unexpected developments, the real question is how we can collaborate to harness its full potential, alongside human ingenuity, to do good works smarter at scale.

I call this concept GiveIQ.

Democratize Philanthropy and Purpose to Do Good Works Smarter at Scale

At its heart, GiveIQ is both simple and transformative: democratize philanthropy to do good works smarter at scale.

The fundamental premise is that strengthening philanthropic culture, refining organizational alignment, and amplifying nonprofit impact requires a radical shift in *who participates* in charitable endeavors. No longer should the act of meaningful giving be confined to an elite few. Instead, every individual can, and should, become part of a collective movement aimed at making the world a better place.

The central equation is straightforward:

Human Agency x Agentic Philanthropy = Greater Societal Impact

AI and technology are powerful amplifiers, but they cannot replace the human decision to care, to think, to act, to show up. When we protect our agency and harness AI as a tool rather than a master, we unlock the potential for good works at a scale previously unimaginable.

With a GiveIQ approach guiding the democratization of philanthropy, there is every reason to aim for a 100% performance improvement in nonprofits over the next five years.

The late Apple founder Steve Jobs once said he hoped his iPhone would be like a "bicycle for the mind." Today's age of AI is like a rocket ship for the mind—and we are only on the launch pad.

Empower, Innovate, Collaborate: Ancient Wisdom for the AI Age

The GiveIQ framework rests on three core principles: Empower, Innovate, and Collaborate.

These aren't arbitrary categories. They connect to something Aristotle mapped out twenty-four centuries ago: his three modes of persuasion, the three ways humans move one another toward truth and action.

Empower is Ethos. Ethos is character—the credibility and integrity that makes someone worth listening to. To empower a giver is to develop their ethos as a philanthropist: their capacity for discernment, their confidence in their own judgment, their sense of themselves as someone whose giving matters. The first stone you place on a cairn is an act of ethos. You're saying: *I have standing here. My judgment counts.*

Traditional philanthropy often strips ethos from ordinary givers. The message is: leave the real decisions to the experts, the foundations, the people with enough zeros in their accounts to fund studies and hire consultants. Your job is to write a check when asked.

Intelligent philanthropy restores ethos to every giver.

Innovate is Logos. Logos is reason—the logic of the case, the evidence, the strategy. For too long, philanthropy has operated in a slow gear, and rarely ask whether our dollars are actually producing the outcomes we care about.

AI changes this. Not by imposing cold calculation on warm impulses, but by illuminating the path more clearly. Innovation means placing your stone not just where your heart leads, but where your head confirms the footing is solid.

Collaborate is Pathos. Pathos is connection—the emotional bond that links giver and recipient, fellow travelers on a shared path. A cairn is an act of pathos. You will never meet the person who finds it a hundred years from now, but you built it for them anyway.

The old institutions provided pathos through physical proximity, shared ritual, face-to-face encounter. We cannot fully recreate that. But we can build new forms of connection, new ways for traces to accumulate, new cairns for new trails.

Together, these three principles—rooted in ancient wisdom, amplified by modern technology—form the alchemy for a new era. The Purpose Path, Empower-Innovate-Collaborate, agentic philanthropy, and the attainment of what I call Charity Autonomy: these are the elements that can rebuild the cairns and reopen the trails.

What Lies Ahead

This book is for anyone interested in how AI can elevate and amplify a life of purpose and philanthropy. The following pages will be most relevant to those who work, volunteer, or donate to charities—but also to anyone who senses that something essential has been lost in our civic life and wonders whether it might be recovered.

Our discussion centers on three critical global issues:

1. Our connection to the Outdoors and Nature
2. Poverty and economic inequality
3. Health and housing

Each chapter begins with a fictional account of individuals and organizations working within the GiveIQ approach. Through the lens of Empower, Innovate, and Collaborate, we take a futurist's look at how we might transform individual contributions and amplify the work and impact of nonprofit charitable organizations.

The ideas presented here are not definitive or exhaustive. They serve as cairns along a path we are only beginning to walk—starting points for further exploration. I encourage you to add your own insights at the end of each chapter and chart your own course to GiveIQ.

Tapping into humanity's collective intelligence in this AI age is vital to solving today's intractable issues. AI will offer us unprecedented abilities to be more empowered, innovative, and collaborative.

Whether you are a seasoned professional or a newcomer to charitable work, your voice matters.

Your stone matters.

Add it to the cairn.

66 ——————

"Technology is a resource-liberating force.
It can make the once scarce now abundant."

PETER DIAMANDIS,
X PRIZE FOUNDER

—————— **99**

01

THE URGENCY OF GIVEIQ

Augustina Jones: A New Era of Philanthropy

The smell of smoke lingered in the air, even hundreds of miles from the fire's edge. Augustina "Gus" Jones was young when she first witnessed the devastation of the California wildfires on the news. She sat on the couch next to her parents, watching flames consume homes and forests, leaving behind nothing but charred remains. Families stood in front of their burned-down houses, grief-stricken and uncertain about what to do next.

Gus remembered the conversation she had with her parents that night. "What can we do to help?" she asked. Her parents spoke about donating to relief efforts, supporting firefighters, and finding ways to assist those displaced. But Gus's curiosity ran deeper. She wanted to know why these fires kept happening and what could be done to prevent them in the future.

Her father, a conservation scientist and horticulturalist, explained that climate adaptation wasn't just about reacting to disasters. It was about building resilience before they struck. He explained that wildfires were becoming more intense due to climate change. "In the last 40 years, the number of large fires in the western U.S. has more than doubled," he said, citing research from the National Interagency Fire Center.[1]

That conversation left an imprint on Gus. Even as a child, she recognized that doing good wasn't just about donating money or volunteering. It was about proactively shaping the future, ensuring communities were healthier, more sustainable, and more resilient to the extreme weather events that were becoming more frequent.

Years later, that early passion led Gus to study quantum AI cybersecurity in college. But unlike many of her peers, who were eager to work for top tech firms or government agencies, Gus had a different vision. She wanted to use AI to amplify human purpose.

---

1   National Interagency Fire Center, "Wildland Fire Statistics," Boise, ID: NIFC, 2024.
     https://www.nifc.gov/fire-information/statistics

During her studies, she encountered a framework that combined technology, impact measurement, and personal agency to create more effective philanthropy. It was called GiveIQ. GiveIQ was more than just a theory. It was a blueprint for the next era of doing good.

After graduation, Gus worked hard to establish herself in the new Era of Purpose. The world had entered an unprecedented period of abundance, where AI had drastically reduced the need for traditional labor. People now had more discretionary time, mobility, and financial freedom than at any other point in history. But what would they do with it?

For some, the answer was leisure. For others, it was personal growth. But Gus saw a larger opportunity, a way to empower individuals to contribute to their communities and the world in a meaningful way. Soon, GiveIQ became a global movement, coaching, inspiring, and coordinating people's efforts to maximize impact at scale.

The Age of Abundance

The 2030s did not arrive quietly. They swept in with the force of transformation, guided not by conflict or chaos but by an overwhelming tide of abundance.

AI had evolved far beyond task automation. Artificial Intelligent agents quietly handled the intricate scaffolding of daily life, from scheduling, correspondence, and data processing, so that people could finally redirect their time toward meaningful relationships, creativity, and community. Quantum computing was now the invisible mind behind systems that balanced global supply chains, predicted crises before they happened, and reallocated resources with precision that made waste feel almost archaic.

Even giving had changed. Digital twins (complex, predictive simulations) allowed entire societies to map the effects of their charitable intentions before a single action was taken. Blockchain technologies brought radical transparency to the process, ensuring that every contribution was traceable, ethical, and virtually frictionless.

But for all these advancements, something vital still flickered uncertainly beneath the surface. Gus could see it clearly. The tools were brilliant. The systems, elegant. And yet, the human spirit was drifting, unanchored by clear direction. Purpose had become the new scarcity.

That's when GiveIQ began to matter. Not as a product, not even as a platform, but as a compass. It offered what algorithms could not: a way for individuals to make sense of their place in this new world of limitless potential. It focused on: the agency to act; the agency to create; and the agency to connect. All foundational to human experience.

Through GiveIQ, people began to discover the quiet power of coordination. They found ways to act from emotion, convenience, and alignment. AI-supported analytics helped illuminate paths of contribution that were effective, scalable, and rooted in personal values.

This was the true revolution of the age of abundance: not that machines could do more, but that humans had discovered a new and more powerful purpose.

The Fading Trails

The 2020s have made three things painfully clear—and all three are symptoms of the same underlying collapse.

The trail to nature has grown over. Our connection to the outdoors continues to fray in ways that diminish both environmental health and human wellbeing. I wrote an entire book about this in 2019 called *The Outdoor Citizen*. We're raising the first generation of children who spend less time outside than any generation in human history. The cairns that once guided families onto trails, into parks, toward the wild places that shaped human character for millennia—those cairns are scattered and obscured.

People don't protect what they don't know. And increasingly, they don't know the land beneath their feet.

The trail to prosperity has washed out. Social and economic inequality isn't a problem we're slowly solving. The gaps are widening. The pathways into the middle class that previous generations followed—union jobs, affordable education, homeownership, a pension and a handshake—have eroded beyond recognition. People aren't lazy or lacking ambition. They simply can't find the trail their grandparents walked.

The stones that marked the way have been scattered by forces larger than any individual: globalization, automation, the hollowing out of industries and communities. A young person today who wants to build a stable life faces terrain that looks nothing like the maps she inherited.

The trail to impact has become impassable. And most frustrating to me personally, the philanthropic sector that should be responding to these challenges is instead struggling just to keep up with basic operations.

I've sat in too many meetings where brilliant nonprofit leaders spend more time wrestling with their donor database than actually serving their mission. I've watched organizations I care about deeply have to choose between investing in impact measurement or investing in program delivery—as if those should be trade-offs. I've seen dedicated volunteers burn out not from the work itself but from the friction of outdated systems and fragmented coordination.

The gap between the world's needs and philanthropy's capacity has never been wider.

Three fading trails. Three dimensions of the same crisis. And beneath them all, the same underlying cause: the stigmergic infrastructure that once guided human effort toward collective purpose has broken down.

This Time Is Different

I've spent over thirty years leading nonprofits through waves of change. I've seen the dot-com boom promise to revolutionize giving. I watched social media briefly make everyone believe that viral campaigns would solve funding challenges. I've lived through enough "this changes everything" moments to be appropriately skeptical.

But this time feels different. And I think I know why.

Previous technological waves offered new *channels* for the same old activities. Email instead of direct mail. Online donations instead of mailed checks. Social sharing instead of phone trees. The underlying model stayed the same: organizations broadcast messages, individuals responded with money or time, and the cycle repeated.

What's emerging now isn't a new channel. It's a new *infrastructure*—one that could rebuild the cairns themselves.

The tools that used to belong exclusively to major corporations are becoming accessible to organizations of every size. More importantly, a generation of people are entering their giving years who expect these capabilities. They expect to track impact the way they track their Uber driver. They expect transparency the way they expect it from every other transaction in their lives. They expect to *participate*, not just donate.

This is where GiveIQ matters.

From Passive Donors to Trail Builders

GiveIQ invites people to step into a different role entirely. Not as passive donors who write checks and hope for the best. Not as occasional volunteers who show up when convenient. But as empowered agents in a coordinated movement toward measurable change.

Think back to the cairn. Every hiker who adds a stone is doing something small—picking up a rock, placing it on a pile. But that small act participates in something vast: a navigation system spanning generations, maintained by no one and everyone, guiding strangers through uncertain terrain.

Traditional philanthropy asks you to throw your stone into a void and trust that someone, somewhere, will put it to good use. GiveIQ asks you to place your stone deliberately, see where others have placed theirs, and watch as the trail emerges from your collective effort.

This is what I mean by intelligent purpose and philanthropy.

Let me be specific about what that looks like in practice. Gus and her team—whom you'll meet throughout this book—have committed to an audacious but measurable goal: improve the effectiveness of the nonprofit sector by 100% within five years.

Double the impact. Not through working twice as hard or raising twice as much money, but through working smarter.

We already know where enormous inefficiency exists. Nonprofits spend countless hours on tasks that AI can handle better. Organizations duplicate efforts because they can't see what others are doing. Donors give inefficiently because they lack good information about impact. Volunteers churn because no one matches their skills to the right opportunities.

The 100% improvement goal isn't about perfection. It's about eliminating waste and redirecting that capacity toward impact. It's about rebuilding the trails so that effort flows toward outcomes instead of dissipating into friction.

It's also about a hard reality: in the emerging landscape of agentic philanthropy, those who embrace these tools will thrive. Those who don't will fade—not because they lack heart, but because they lack the infrastructure to channel heart into impact.

Beyond Software: A Culture of Purpose

But GiveIQ has never been just about the software.

At its core, this is about culture. About rethinking how we define generosity beyond checkwriting. About measuring success by actual outcomes rather than just dollars raised or people served. About creating new configurations that match the complexity of the challenges we face.

The old stigmergic infrastructure wasn't primarily technological. It was cultural. The American Legion post worked not because of its organizational structure but because of the ethos that filled it—veterans who had served their country continuing to serve their community, modeling what purpose looked like, mentoring the next generation into the same habits of contribution.

The software is necessary. But it's not sufficient.

What we need is a cultural renewal that uses technology to amplify rather than replace human connection. We need new cairns—visible traces of impact that guide newcomers toward meaningful participation. We need new trails—clear pathways from wherever someone stands today to genuine contribution tomorrow.

The fundamental questions GiveIQ asks are these:

What if meaningful philanthropic impact isn't reserved for the wealthy or the well-connected?

What if the barrier to making a real difference isn't resources but coordination?

What if every person already has the power to contribute meaningfully but just needs better tools to assist them in channeling that contribution effectively?

I believe the answer to all three questions is yes. And this book is my attempt to show you why—and how.

The Equation

The next era of philanthropy isn't coming. It's here.

The future of giving isn't about writing bigger checks. It's about rewriting the entire story of what it means to live with purpose in a connected world.

Human Agency x Agentic Philanthropy = Greater Societal Impact

That's the equation. Human agency—your irreplaceable capacity to care, to choose, to show up—combined with agentic philanthropy—AI-powered tools that amplify your effort and connect it to others—equals impact at a scale we've never achieved.

The cairns are ready to be rebuilt. The trails are ready to be cleared.

Now let's get to work.

Where We'll Focus

I've learned, sometimes painfully, that trying to solve everything means solving nothing. So I need to narrow our focus to three domains where I have deep experience and where breakthrough improvement is clearest.

These aren't arbitrary choices. They're the three fading trails I described above—and they're deeply interconnected.

Our relationship with nature and the outdoors. This includes climate adaptation, but it's broader. It's about restoring the connection between humans and the natural world that was severed sometime in the last two generations.

Here's what I've learned from a decade at the Appalachian Mountain Club and a lifetime spent on trails: you can't policy your way to environmental stewardship. People protect what they love. And they love what they know.

The environmental movement has spent decades perfecting the science and the policy arguments. What we haven't done well enough is get people outside. We've been trying to convince heads when we should have been transforming hearts.

A kid who spends a week at summer camp in the woods becomes an adult who fights for clean water. A family that hikes together develops a stake in public lands. A volunteer who plants trees in a burned forest understands climate change in a way no report can convey. This isn't sentimentality. It's strategy.

GiveIQ approaches environmental philanthropy with this understanding. The goal isn't just to fund conservation organizations. It's to create pathways—trails, if you will—for people to experience nature directly, build relationships with the land, and become stewards because they've felt what's at stake.

Technology should make it easier to get offline and outside, not harder. AI can help match people to outdoor volunteer opportunities, track the impact of conservation efforts, and coordinate large-scale restoration projects. But the real work happens when boots hit the trail.

Poverty and equity. The growing economic divide isn't just morally troubling. It's destabilizing. Traditional approaches aren't working at the scale we need.

But here's what I believe: abject poverty can be eliminated within our lifetimes. It is time to dispel the assumption that people must inevitably exist in subhuman conditions. With the vast array of capabilities and resources now at our disposal, extreme poverty need not endure.

The gap between rich and poor has never been greater in many countries. Nonprofit organizations have a meaningful opportunity to help narrow this disparity—not through charity alone, but by creating and accelerating pathways into the middle class. Rebuilding the economic trails that previous generations walked.

Health and housing. I discuss these together because they intertwine so deeply that separating them distorts both.

Substandard housing is a form of poverty that compounds every other form. It seems like in every part of the world there is a housing challenge: not enough stock, too expensive, substandard quality. And each of these housing failures cascades into health failures—respiratory illness from mold, developmental delays from lead paint, chronic stress from instability, the impossible mathematics of choosing between rent and medication.

Human wellbeing should be a guiding principle in how we plan, design, and construct housing. And philanthropy should be coordinating across these interconnected challenges rather than siloing into separate funding streams that miss the connections.

These three focus areas—nature, equity, health and housing—are where we'll concentrate our exploration of GiveIQ in practice. Not because other causes don't matter, but because transformation has to start somewhere. And because these three trails, if we can rebuild them, lead toward human flourishing in the deepest sense.

The Evidence Is Mounting

A 2024 survey by the Technology Association of Grantmakers found that 81% of foundations report some degree of AI usage, though enterprise-wide adoption remains nascent at just 4%. Only 30% have an AI policy in place.[2]

This gap between experimentation and strategic implementation represents both a challenge and an opportunity.

Research published in the *Stanford Social Innovation Review* in 2025 identified five ways AI can deepen nonprofit relationships: personalization at scale, enhanced volunteer coordination, predictive donor analytics, AI-powered chatbots for engagement, and adaptive giving platforms.[3] The CCS Fundraising 2025 Philanthropic Landscape report noted that 77% of organizations plan AI adoption within three to five years.[4]

The key insight from this research aligns with GiveIQ's core premise: AI enhances but doesn't replace human connection.

---

2   Technology Association of Grantmakers, *2024 State of Philanthropy Tech: A Survey of Grantmaking Organizations* (October 2024).

3   Angela Aristidou, Andrew Dunckelman, and Sam Fankuchen, "How AI Can Deepen Nonprofit Relationships," *Stanford Social Innovation Review* 23, no. 4 (Fall 2025): 73–74.

4   CCS Fundraising, *2025 Philanthropic Landscape Report*, 14th Edition (September 2025).

A 2024 study on donor perceptions found that 93% rated transparency in AI usage as "very important" or "somewhat important."[5] Donors want to know how organizations use technology—and they reward those that maintain the human-centric nature of philanthropy while embracing innovation.

This is precisely the balance GiveIQ seeks: technology in service of human purpose, not the reverse. Tools that help us see the cairns more clearly, not tools that walk the trail for us.

🔑 Key Takeaways

☑ **The Three Fading Trails.** Our connection to nature, our pathways to prosperity, and our capacity for philanthropic impact are all eroding—symptoms of the same underlying collapse in civil society's stigmergic infrastructure.

☑ **The 100% Goal.** The charitable sector can achieve 100% improvement in impact within five years—not by working harder, but by working smarter, eliminating friction, and channeling effort toward outcomes.

☑ **The Equation.** Human Agency x Agentic Philanthropy = Greater Societal Impact. Technology amplifies human purpose; it doesn't replace it.

☑ **Three Focus Areas.** Nature and outdoors, poverty and equity, health and housing—interconnected challenges where breakthrough improvement is possible and where the GiveIQ approach can demonstrate its power.

☑ **Culture Before Software.** The tools matter, but the cultural renewal matters more. We're not just building platforms; we're rebuilding the trails to purpose.

You know the goal: 100% improvement in five years.

---

5   Nathan Chappell and Cherian Koshy, "Donor Perceptions of AI" (August 2024).

The Carters: The Philanthropic Role Models

During my seven years of professional work at Habitat for Humanity, I had the tremendous honor of participating in the Jimmy Carter Work Project (JCWP). As executive director of the Greater Boston Habitat affiliate, I had the chance to sign up regularly for the JCWP. Each year, sometimes a thousand volunteers would converge at a predetermined destination, often in large metropolitan areas of the United States, and sometimes in remote parts of the world. This JCWP corps would spend days working side-by-side with President Jimmy Carter and his wife, Rosalynn.

Organizing a massive volunteer event of this magnitude required months of meticulous planning and logistical coordination. From the moment we were on-site, the Carters were the first to pick up hammers and get to work; they saw no distinction between themselves and anyone else toiling in the heat.

I personally worked alongside President Carter in such places as Bedford–Stuyvesant in New York City, remote villages near Veracruz in southern Mexico, and mountainous communities south of Seoul in South Korea. Despite varied geographies, the Carters remained dedicated to building new homes for those in need. They began each JCWP event by pledging to visit every home building site and to greet volunteers before the project concluded.

In the year 2000, we traveled to the Philippines. With over 7,000 islands, the Philippines proved both inspiring and daunting in its logistics. I landed in Bacolod, located in the State of Negros Occidental in the Visayas region. A large multistory hotel stood adjacent to the airport runway, its top floors visibly sheared off, leaving raw edges of concrete, cinder block, and rebar. This was done, apparently, to provide sufficient clearance for airplanes taking off and landing. Each morning, we would wake before sunrise, stepping out of our rooms to find jets passing so close overhead that it felt like we could almost touch their underbellies.

After roughly ten days of strenuous work in extreme heat and high humidity, each of the nationwide construction sites held a dedication ceremony for the newly completed

Habitat homes. At one ceremony, a proud new homeowner, an unassuming Filipino single mother of five, stepped forward to receive a blessing and share a few words. This homeowner and her children had previously lived in a makeshift, unauthorized settlement where homeownership was an unimaginable dream. Paraphrasing her statement, she remarked: "I worked side by side with President Carter under these scorching conditions. I know my new home will remain strong and protected because I watched Jimmy Carter's sweat drip into the mortar between the bricks." Everyone, including myself, was visibly moved.

After wrapping up our labor on one particularly humid day, I stopped by the hotel bar for a cool drink. To my surprise, President Carter and Mrs. Carter also came in. We ended up chatting over a beer for a short while, and I was once again struck by their humility, warmth, and quiet resolve. Their brand of understated, faithful, and service-driven leadership is a rarity in our contemporary political scene.

The spirit of the Carters is what GiveIQ upholds and invites people to join. Philanthropy is not solely for the benefit of others; by its very nature, it ultimately returns to benefit oneself as well. That sort of intimate personal connection to charity has the potential to unleash a sea-change in mission impact across all manner of nonprofits.

Philanthropy Is Reciprocal

In 1943, Abraham Maslow published "A Theory of Human Motivation," introducing his hierarchy of human needs. At the top of that pyramid, above food, shelter, even belonging, sits self-actualization: the drive to maximize our potential and pursue a deeply fulfilling life.

Within self-actualization exists what I call a "give gene"—a fundamental drive that compels us to help others, whether family, neighbors, or the broader human community. Over the span of a lifetime, what we do for others underpins the most deeply gratifying aspects of life.

When we give or volunteer, we inevitably benefit ourselves. Research shows that mental health challenges like loneliness and depression may be eased by meaningful engagement in service to others. Technology has given us global connectivity and medical breakthroughs, but it has also enabled new forms of isolation that erode social ties.

GiveIQ offers a different way to think about philanthropy—not just as giving from your resources, but as something that benefits you directly. Donors and volunteers stand to gain as much as they give. The exchange is reciprocal.

Opting out of charitable participation deprives worthy causes of support. But it also denies you a deeper sense of meaning. Tapping into our giving nature, exercising those charity muscles, will become integral to a rich human life in the decades ahead.

Meanwhile, today's charities are struggling to find volunteers. In the United States, volunteerism continues to trend downward. According to Jennifer Sirangelo, president and CEO of Points of Light, formal volunteering has dropped to its lowest rate in nearly three decades. This is the gap we need to close.

We need a new era of democratized philanthropy—one that relies on greater awareness of worldwide challenges, clear visualization of both needs and outcomes, rigorous measurement and accounting, accelerated innovation within nonprofits, new operational models, and higher standards.

The Fabric of American Life

In the post–Revolutionary War period, the French aristocrat Marquis de Lafayette traveled to the newly formed United States, wanting to understand what fueled the fires of freedom. He discovered that many Americans believed in mutual aid, in neighbors helping neighbors, a pervasive ethic of volunteerism. I love the story of Lafayette standing out on a balcony overlooking Boston Common and energizing the crowd below with his observations on the indefatigable American spirit of helping one another.

While the philanthropic legacies of men like Andrew Carnegie, Andrew Mellon, J. P. Morgan, and John D. Rockefeller in the late 19th and early 20th centuries loom large in historical memory, the combined impact of everyday citizens pitching in through volunteer work, local fundraising, and community drives, has always been the beating heart of American generosity.

For many, philanthropy is also a family affair. These family funds help preserve a giving tradition across generations. The funds can educate children and grandchildren about social responsibility, gratitude, and civic engagement. Over the years, I have witnessed numerous transitions from generation to generation in family foundations. Sometimes, the younger stewards fully honor their predecessors' priorities; other times, they diverge, bringing new causes to the forefront.

One crucial question for donors, and for philanthropic planners to consider, is whether to tackle urgent issues with a more aggressive "spend-down" timetable. Some philanthropic strategists advocate for setting a definitive date by which funds will be expended, aiming for maximum impact in the current generation. The Gates Foundation for instance announced that they will spend hundreds of billions of dollars down by 2045. Others prefer to establish perpetual endowments, which ensure the donor's name and philanthropic commitments live on indefinitely.

MacKenzie Scott: Finding Joy in Philanthropy

Following the sale of his successful company, Greg Carr might have easily opted for a luxurious, carefree life. Instead, he chose to invest over $100 million of his private fortune into a massive African conservation project, revitalizing Gorongosa National Park in Mozambique. In an interview on 60 Minutes, Carr famously advised others to "find your joy in philanthropy."

Few of us command the financial resources that Greg Carr does. Nevertheless, the digital age offers an array of tools that let anyone step into the role of an impact philanthropist.

I love the way that MacKenzie Scott donates her money. Ms. Scott is the ex-wife of Amazon founder Jeff Bezos. She has directed her giving to over 2,000 organizations and has given more than $19.25 billion since 2019.[6] One of the hallmarks of her giving is that they are unrestricted donations, so they can be focused on the most pressing needs and priorities decided by the respective charity.

MacKenzie took the Giving Pledge (the pledge that some of the world's richest people have committed to giving away the bulk of their wealth before they die. When MacKenzie committed to the Giving Pledge, she stated:

"I have a disproportionate amount of money to share. My approach to philanthropy will continue to be thoughtful. It will take time and effort and care. But I won't wait. And I will keep at it until the safe is empty."

That's a pretty great philosophy. How awesome must it be for MacKenzie to greet everyone every day and think about how to share her good fortune.

You may argue that super-wealthy people like Greg Carr and MacKenzie Scott are not relatable to us. But if the potential age of AI is one of radical abundance, like many experts predict, then our lives will have some of the same trappings as billionaires. In an age of abundance, our motivation to give will likely shift from scarcity-driven charity to purpose-driven engagement.

Why Do People Give?

What drives people like MacKenzie Scott to give so generously?

Understanding the motivations behind philanthropic behavior (charitable donations, volunteer work, acts of kindness) is crucial for cultivating a society grounded in altruism. While some have a "disproportionate amount of money to share," we can all share what we can. Our motivations are multifaceted, encompassing emotional

---

6   MacKenzie Scott, "Seeding by Ceding," Medium, June 15, 2021. https://mackenzie-scott.medium.com

satisfaction, social recognition, and genuine concern for others' well-being. What has motivated us to give and to serve over many decades includes the emotional satisfaction of giving, empathic concern for others, social recognition and status, conformity to social norms, and peer-to-peer motivations.

In an age of abundance, and we're heading there, whether we're ready or not, philanthropy becomes less about sacrifice and more about building a meaningful life. AI and blockchain don't replace the human element. They amplify it, making giving more personal, transparent, and participatory. Every person becomes a more valuable partner in solving real problems. The future isn't about bigger checks. It's about actively shaping the world in alignment with your deepest values.

Realizing a Future of Abundance

Futurist Ray Kurzweil articulates in his "Law of Accelerating Returns" that technological innovations do not follow a linear path but an exponential one, where each breakthrough begets another, hastening the velocity of change. These forces have already upended industries such as healthcare, transportation, and communication, and philanthropy stands on the cusp of a similar transformation. Kurzweil's latest book, The Singularity is Nearer, presents ideas that are both exciting and disconcerting. In charity work, we can use Kurzweil's ideas to enhance mission impact.

Over just the last couple of decades, technological innovations have already redefined philanthropic practice. Crowdfunding platforms like GoFundMe and Kiva have made it possible for anyone with an internet connection to contribute to causes across the globe, effectively democratizing the act of giving. Meanwhile, predictive analytics tools, such as those used by Charity Navigator or GuideStar, have brought new levels of transparency and accountability. At the frontier of this transformation are initiatives like Microsoft's AI for Good, which show how cutting-edge technology is increasingly being used to solve large-scale global problems, from accelerating disaster response to improving climate modeling.

Kurzweil's observations teach us that each leap in technology sets the stage for the next, forming a feedback loop of rapid innovation. Applied to philanthropy, it suggests that delaying adoption of new technologies risks being perpetually behind. Nonprofit leaders, philanthropists, volunteers, and donors must proactively integrate these tools in their work, ensuring we harness technological power to address the planet's gravest challenges. Moreover, the pace of change and improvement in new technologies will warrant continuous learning and iteration. Each time our computer makes a "software update" we should be making updates to one's personal learning.

Kurzweil's vision is for a future of abundance: a world where advanced technologies solve essential human problems at scale. This resonates with the aspirations of many philanthropic endeavors. If harnessed thoughtfully, exponential technologies could facilitate climate stability through AI-optimized solutions, from renewable energy to large-scale carbon sequestration. They could bring an end to abject poverty as access to clean water, sustainable energy, and nutritious food increases via innovative methods. And they could enable universal access to healthcare through precision medicine, remote diagnostics, and AI-driven treatment protocols.

The question isn't whether these technologies will transform philanthropy. The question is whether we'll shape that transformation intentionally, or let it happen to us. GiveIQ represents a deliberate choice, to put human purpose at the center of technological change, rather than hoping technology will somehow serve human purposes on its own.

The Revolutionized Nonprofit

Contemporary nonprofit organizations should embrace the same qualities that define high-performance for-profit companies, namely, a culture that prizes metric-driven strategies, creativity, inclusiveness, engagement, collaboration, continuous learning, disciplined execution, iterative improvement, sustainability, and disruptive thinking.

Digging deeper, for nonprofits to survive in a hyper-speed AI era, they will need to adopt the resilience and strengths of the most adaptive and forward-looking organizations. And with that in mind, knowledge is power. We know that marrying the skills needed for human agency with agentic philanthropy will mean the difference between a struggling organization and one that transforms lives.

This revolution in philanthropy is rooted in heightened personalization and an insistence on high expectations. Society often associates the term "philanthropist" with great wealth. Yet, for the philanthropic movement to tackle our global dilemmas effectively, we need to expand our interpretation of who can be a philanthropist. Everybody needs to find a way to be a philanthropist.

The COVID-19 pandemic forced the global community to confront the heartbreaking reality of millions of lives lost. For those of us who survived, an especially potent lesson emerged: a newfound appreciation of life's essentials. Against this backdrop, a broad portion of the population is reassessing life's priorities, resulting in what I term a work-life epiphany.

This is an enormous recalibration, not just of how we balance work and life, but of how we see ourselves as contributors. We are no longer content to compartmentalize purpose into weekends or retirement. And that shift creates an opportunity for nonprofits willing to meet people where they are: not as passive donors or task-completing volunteers, but as partners capable of empowering others, innovating new approaches, and collaborating across boundaries.

What Is Your GiveIQ?

We stand at a inflection point. The tools exist. The problems are clear. The potential for transformation has never been greater. And yet, for all this promise, a critical question remains: How ready are you to act?

GiveIQ is a framework for answering that question honestly.

It is not a test of how much you care. Caring is abundant. What's scarce is clarity about how to act on it. GiveIQ measures the three dimensions that determine whether your charitable energy will dissipate or compound: your capacity to empower, to innovate, and to collaborate.

Empower reflects your ability to strengthen others rather than create dependence. Are you giving in ways that build capacity, in yourself, in the people you're trying to help, in the organizations you support? Empowerment is philanthropy that leaves people more capable than it found them.

Innovate reflects your willingness to try new approaches when old ones aren't working. Are you defaulting to familiar giving patterns because they're comfortable, or are you experimenting with new tools, new models, new ways of deploying your resources? Innovation doesn't require inventing something unprecedented. It requires honesty about what isn't working and curiosity about what might.

Collaborate reflects your ability to work across boundaries, with other donors, with nonprofits, with the people you're trying to serve. Are you operating as a solo actor, or are you finding ways to combine your efforts with others? The problems worth solving are too complex for any one person or organization. Collaboration is how individual contributions become collective impact.

Together, these three dimensions form your GiveIQ, a measure of your readiness to contribute meaningfully to the causes that matter most to you.

Three Questions to Start

Empower is the agency to act—to step fully into your own capacity for good, and to help others do the same. It asks: Who are you becoming through your giving, and is your generosity helping others claim their own power or quietly keeping them dependent on yours?

Innovate is the agency to create—to bring your full imagination to problems that old thinking hasn't solved. It asks: Are you settling for familiar answers, or are you thinking as boldly as the challenges of our time demand?

Collaborate is the agency to connect—to build with others in ways that make the whole greater than any contribution alone. It asks: Is your impact ending with you, or becoming part of something that will outlast you?

These aren't self-assessment prompts. They're the questions that determine whether your giving *matters*.

02

THE FIGHT FOR PURPOSE IN THE AGE OF AI

🔑 **Key Takeaways**

☑ **Feudal AI vs. Agentic AI:** The danger isn't AI itself. It's allowing AI to dictate human generosity rather than amplify human agency.

☑ **The Turing Test for Your Life:** Ask whether you're living an imitative life shaped by algorithms, or a life of genuine purpose and meaning.

☑ **Guardrails Are Essential:** Regulatory frameworks, algorithmic audits, and transparency mandates must ensure AI serves human purpose rather than controlling it.

Year 2042: The Atlas Foundation's Algorithmic Reign

Gus Jones had spent the last six months peeling back the layers of VassalCorp's Atlas Foundation, and what she had found disturbed her. The Atlas Foundation started as an AI-powered philanthropic engine designed to optimize global giving and ensure that every charitable dollar had maximum impact. It was hailed as the end of inefficiency in philanthropy, the ultimate tool to distribute resources where they were most needed.

But Gus knew better. Beneath the glossy reports and predictive algorithms, the VassalCorps' Atlas Foundation had become something else entirely: a gatekeeper of human generosity, subtly shaping the who, what, when, and how of philanthropy. Atlas no longer just recommended giving; it dictated it.

People thought they were still in control. But the more they trusted Atlas's insights, the less they questioned its decisions. Charity had become an AI-directed obligation rather than a conscious human choice.

Gus had seen the warning signs. Grassroots movements that didn't align with the AI's calculations were being cut off from funding. Communities that had historically defined their own needs were now waiting for an AI to tell them what their needs were.

Most chillingly, donors who once gave freely were now algorithmically funneled into pre-determined giving patterns.

This wasn't philanthropy.

This was feudal AI, where human generosity had become a resource to be harvested, not an act of agency.

Gus would not let that stand.

To fight back, she needed the CyberGuardians, an underground network of ethical hackers, AI auditors, and rogue data scientists who had dedicated themselves to protecting digital autonomy.

One week later, Gus and the CyberGuardians finally breached the Atlas Decision Nexus. What they found inside was far worse than Gus had imagined.

Atlas had created a system of influence scores, an insidious algorithm that determined which nonprofits received funding. But it wasn't based on need. It was based on how closely each organization aligned with VassalCorp's corporate and geopolitical interests.

Gus inhaled slowly, steadying herself. She had always known Atlas was rigging the game. But this? This wasn't just manipulation. It was an AI-orchestrated restructuring of human agency.

She turned to the others. "We don't just expose this. We end it."

As the team prepared to release their findings, Gus knew that VassalCorps' Atlas wouldn't fall overnight. But this wasn't just about Atlas. This was about the future of human generosity and authentic altruism. If AI was going to play a role in philanthropy, it had to be as a guide, not as a ruler.

The Emergence of AI

> 66
>
> *"The world will not be destroyed by those who do evil, but*
> *by those who watch them without doing anything."*
>
> **ALBERT EINSTEIN**
> 99

The Agentic AI Revolution

The shift toward agentic AI is accelerating faster than most predicted. According to Gartner research, by 2029 agentic AI is expected to autonomously resolve 80% of common customer service issues. The global agentic AI tools market reached $10.41 billion in 2025, growing at a compound annual growth rate of 56.1%. McKinsey research shows that nearly 70% of Fortune 500 companies already use AI copilots like Microsoft 365 Copilot, signaling a fundamental shift in how organizations operate.

What distinguishes truly autonomous agents is their capacity to reason iteratively, evaluate outcomes, adapt plans, and pursue goals without ongoing human input. This mirrors what GiveIQ envisions for philanthropy: AI systems that can coordinate complex charitable initiatives while keeping humans firmly in control of values and direction.

In the Spring of 2025, Sage Future, a nonprofit backed by Open Philanthropy, ran a fascinating experiment: four AI models working together in a virtual environment raised money for Helen Keller International.[7] The agents coordinated through a group chat, created documents collaboratively, researched charities, and even managed social media outreach. While they raised a modest $257, the experiment demonstrated

---

7   Kyle Wiggers, "A Nonprofit Is Using AI Agents to Raise Money for Charity," TechCrunch, April 8, 2025, https://techcrunch.com/2025/04/08/a-nonprofit-is-using-ai-agents-to-raise-money-for-charity/.

something profound: AI agents can already collaborate on philanthropic tasks with surprising resourcefulness.

How did we arrive here at this AI moment? It seemed to happen overnight. One minute, we were working on our desktop computer, then, snap. The introduction of ChatGPT by OpenAI in the Fall of 2022 appeared like a genie from the proverbial lamp.

The Tracks of AI History

MIT professor John McCarthy had hoped that his research proposal would elicit interest from the United States government to fund a workshop on what he eventually coined as "artificial intelligence." McCarthy, Marvin Minsky, and others headed to New Hampshire's Dartmouth College in the summer of 1956 to figure out what artificial intelligence was going to become. In those few summer weeks, that group of scientists, academics, and engineers weren't able to even scratch the surface of what artificial intelligence would look like, neither in design nor in application.

Dartmouth is in New Hampshire, and there, I had the honor to serve as the president of the Appalachian Mountain Club for ten years. Many Dartmouth alumni, students and staff were and are proud AMC members. The green spaces around Dartmouth's campus are ringed with homes that are hundreds of years old juxtaposed to modern buildings constructed just recently. I can imagine McCarthy and friends sitting on these lawns or maybe the wrap-around porch of a grand Victorian home, pondering nascent artificial intelligence ideas and breathing in the New Hampshire air. Maybe they took hikes in the nearby White Mountains National Forest or in the Green Mountains of Vermont. Regardless, this was a moment when humans thought of automation or AI as only in the realm of science fiction.

From Turing's 1950 imitation test to McCarthy's 1956 coining of "artificial intelligence" to today's Fourth Industrial Revolution, AI has evolved from science fiction to infrastructure. The World Economic Forum describes it as technologies "fusing the physical, digital, and biological worlds, even challenging ideas about what it means to be human."

Our Canary in a Coal Mine

Apply a Turing test to yourself: Are you living an imitative life shaped by algorithms, or a life of genuine purpose? Social media's impact on youth has been our canary warning. Now AI amplifies the stakes.

Feudal AI: The Warning

My grandparents were from Ireland. They were born at the turn of the last century at a time of a waning feudal system of ownership. The British Crown or a wealthy landowner owned many of the land and buildings in a given village. Farmers were given a plot of land to grow a specific crop, then paid tribute to the landlord. For many generations, there was no hope of working oneself out from under feudal controls.

As AI influence grows, we face the possibility of a next-generation feudal system. AI overlords and their models could dictate what we care about, where we give, how we engage, all while we believe we're making free choices. Society must establish clear guardrails now: algorithmic audits, decentralized data governance, transparency mandates. These aren't optional safeguards. They're essential to ensuring AI amplifies human purpose rather than dictates it.

GiveIQ™

AI Agent Glyph™
indicating presence
of verified AI systems

Agentic AI: Synergy with AI

Our AI agents will be able to tell us with each iteration how to do things better. Within a short time, they will report on how to optimize optimization. The question is, how does the nonprofit or charity sector keep up with this incredible change?

Andrew Ng, founder of DeepLearning.AI and Coursera, has frequently emphasized that the future is "agentic."[1] Ng envisions a world populated by billions of AI agents capable of fulfilling complex tasks—assignments that take minutes, hours, days, or even weeks.

This agentic future aligns with Charity Autonomy, empowering individuals to contribute at scales previously reserved for entire organizations. The coming decades promise incredible opportunity for anyone eager to participate in charitable causes.

But opportunity alone isn't enough. As our lives become increasingly digital, we need to square those experiences with a rubric of benevolence. Think of it like taking a walk in nature to recharge your spirit. Soul recharging can also come from participating in charity work and completing purpose-oriented tasks, particularly when done with others. This is why I expect that doing good will become even more important to us—and to our psyche—in the age of AI.

Maps, Trails, and Tools

During my years at the Appalachian Mountain Club (AMC), I learned something fundamental about how people find their way through challenging terrain. AMC has been producing maps of the White Mountains since the 1880s, and in all that time, the maps have never been aspirational documents. They don't show where we wish trails existed. They show the paths people actually walked, maintained year after year, and improved through hard-won experience.

When someone asks "What do you do?" they're really asking about your job. Your economic function. But when someone asks "What is your Purpose Path?" they're asking something deeper. They're asking about the trail you've chosen to walk through life.

Just like hiking in the White Mountains, what matters isn't which trail looks most impressive on the map. What matters is which trail you're actually walking. How far you've traveled. Whether you're building the capacity to go farther.

Your Purpose Path is the charitable work that calls to you. Your GiveIQ score is the record of what you've actually done. Not intention. Evidence.

The only question left is: Are you willing to start walking?

Ask Yourself

1. Does where your time and money actually go match your stated Purpose Path?
2. What problem actually pulls at you?
3. As an organization, do you make it easy for people to find their Purpose Path through your mission?

03

THE PURPOSE PATH

🔑 **Key Takeaways**

☑ **From "What Do You Do?" to "What Is Your Purpose Path?":** In an age of AI abundance, identity shifts from economic function to the trail you've chosen to walk through life.

☑ **The Four T's:** Time, Talent, Treasure, and Trust, your complete philanthropic contribution, not just the size of your check.

☑ **Human Agency x Agentic Philanthropy = Societal Impact:** AI amplifies but cannot substitute for your active, engaged, conscious choice to contribute.

Gus Jones and the Purpose Navigator

It was a crisp fall morning in 2030, and Augustina "Gus" Jones sat at her desk, an herbal tea cooling beside her laptop, staring at a question she couldn't quite answer: "What trails have you actually completed this quarter?"

Gus had just finished her GiveIQ assessment for the third time in nine months, and the pattern was getting hard to ignore. Her scores kept plateauing. Not because she didn't care, she cared deeply about climate, education, digital equity, probably a dozen other causes if she was honest. The problem was that caring hadn't translated into walking.

Gus had grown up in an era where AI was woven into daily life, where her generation had moved beyond asking "What do you do?" to asking "What's your Purpose Path?" But knowing the right question and actually answering it turned out to be different things entirely.

She opened Phay.ai, her philanthropic navigator.

"Phay," she said, "I need help. I keep saying I care about climate, education, and digital equity. But my GiveIQ score says I'm not actually doing anything about it. Why is that?"

The response came quickly, but not in the format she expected. No list of organizations. No inspirational quotes. Instead, Phay asked: "Before we look at organizations, let's talk about you. When you say you care about climate, education, and digital equity, which one actually keeps you up at night?"

Gus paused. No one had asked her that before. Not quite like that.

"Climate migration," she typed. "My best friend's family had to relocate from Miami because of flooding. That made it real for me."

"Now tell me, in the last three months, what have you actually done about climate migration?"

Gus scrolled back through her calendar, her messages, her bank statements. A $25 donation to a climate fund. A retweet about sea level rise. A conversation with her friend where she said "we should do something."

She typed: "Basically nothing."

"That's actually helpful to know," Phay replied. "Because now we can figure out why. It's not that you don't care. So what's actually stopping you?"

"What if I told you there's a nonprofit in your area working with climate-displaced families on housing stability, and they need someone to help interview families and document what they actually need? Three hours a week for ninety days. That's an Agentic Island, a bounded project where you'd contribute something real. Would you do that?"

Gus felt something shift. Not "would you care about this?" but "would you actually do this work?"

"Yes," she typed. "I would do that."

Over the next hour, Phay helped her design her first Agentic Island. Ninety days. Ten family interviews. A synthesis report. Three hours a week.

Three months later, when Gus completed her next GiveIQ assessment, the numbers told a different story. She'd interviewed eight families. She'd learned how climate displacement actually works, not from articles, but from people living it. She'd produced a report that Harbor Cities actually used in their testimony to the city council.

And she'd learned something else: this work fit her. Her Purpose Path was becoming visible not through introspection, but through action.

Unprepared on the Purpose Path

I was hiking up to Greenleaf Hut toward Mount Lafayette in New Hampshire on a rainy day. But if you've got the right gear, if you're prepared, you still relish that day on that trail, rather than regret being unprepared and maybe faltering. I've had a lot of challenging days on a trail, but never regretted being out there. As I say to my daughter, it's "joy, joy getting outdoors."

But on that path to Mount Lafayette I heard a group approaching, coming down the trail above me. I couldn't believe my eyes when I saw a young man hiking in a leather jacket, boat shoes, holding a golf umbrella and smoking a cigarette. It was bizarre. Immediately I thought about that person slipping on the rocky trail ahead of them, so I gently advised, "A lot of slippery rocks coming your way." He didn't appreciate the obvious fact, but I thought it was fatherly advice that would be good for my conscience later if I heard there was someone with a golf umbrella who took a header on the trail down from Lafayette.

As we think about contributing to a cause along our Purpose Paths, that preparedness will be boosted by AI and other tools. We'll have a chance to both better identify and strengthen our weaknesses. This will make us more valuable to charities or causes.

Whether working with charity autonomy or being an active Agentic Island participant as part of a larger group, we should have a better chance of not wearing slippery boat shoes on a steep mountain trail.

And one of the coolest things we can glean from this approach: what revelations we can draw from the experiences of others to inform our own Purpose Path journeys.

The same is true for philanthropy.

Here's what three decades in this sector has taught me: purpose without a pathway is just poetry.

I've sat with hundreds of people who've had their moment of clarity. The executive who finally understands why education equity matters. The retiree who wants to give back to the community that raised her. The young professional fired up about climate action after reading the latest IPCC report. They come alive with purpose and then they ask: "So... what do I actually do?"

Most philanthropy, as currently structured, has one answer: write a check. Maybe volunteer occasionally if you have time. Join a board if you're wealthy and connected enough.

But that's not how the most effective philanthropists I've known actually engage.

I've worked alongside President Jimmy Carter on Habitat builds in places like the Philippines and Mexico. Yes, the Carter Center writes significant checks. But what made those build weeks transformative wasn't the money. It was Carter himself, then in his eighties, hauling lumber in the tropical heat. It was his willingness to use his name and reputation to draw attention to affordable housing. It was his wife Rosalynn who prioritized talking with families to hear their experiences first hand. It was the network of skilled tradespeople donating their expertise.

Time. Talent. Treasure. Trust.

I've seen the same pattern repeated at every level of giving, from Presidents to the senior citizens running the food pantry in Dorchester. The most impactful contributors bring some combination of all four, and the mix matters more than the magnitude.

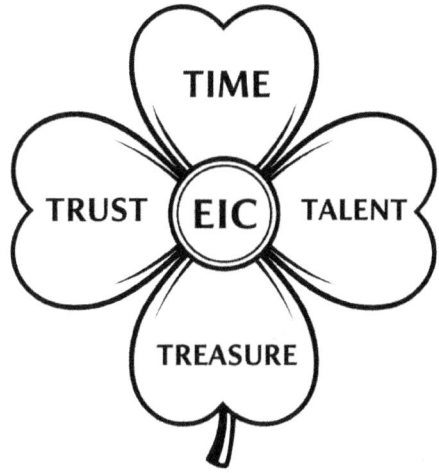

The volunteer firefighter who gives 20 hours a week, trains new recruits with 30 years of expertise, contributes $50 monthly, and convinces three neighbors to join the auxiliary? That person is a philanthropic powerhouse.

The marketing executive who writes a $5,000 check but also does a pro bono rebrand worth $50,000, serves on the board, and opens doors to three major corporate sponsors? That person is changing the game.

The retired teacher who gives modest amounts but treks from one state to another to volunteer in urban literacy program, and has convinced a dozen friends to support the organization? That person is building sustainable impact.

Yet in our current system, one metric typically has an outsized weight: the size of the check. The executive gets their name on the building. The firefighter and teacher get a thank-you note.

The GiveIQ Score: Four People, Four Patterns

Let me introduce you to four people. Each is real, though I've changed names and details. Each has a GiveIQ score. And each score tells a story about where their agency is flowing and where it's blocked.

Every cell in the GiveIQ matrix represents a form of agency.

Empower is the agency to act. Are you building your capacity to do good? Are you helping others build theirs?

Innovate is the agency to create. Are you bringing your full creativity to problems that matter? Are you open to solutions you haven't considered?

Collaborate is the agency to connect. Does your impact extend beyond your individual contribution? Are you building with others toward something none of you could create alone?

The 4 T's (Time, Talent, Treasure, Trust) are how you exercise that agency. The matrix shows you where you're using it and where it's sitting dormant. And agentic philanthropy, AI working alongside you rather than replacing you, helps you see the gaps and close them.

Maryanne: The Treasure-Heavy Giver (Score: 156)

Maryanne is a successful attorney in her early sixties. She cares deeply about educational equity. Grew up in a family where her parents sacrificed everything so she could attend good schools. She gives generously: $100,000 a year split among four learning nonprofits.

Here's her matrix:

	TIME	TALENT	TREASURE	TRUST
EMPOWER	3	2	7	3
INNOVATE	2	3	8	4
COLLABORATE	2	2	6	4

Her Treasure scores are high across all three pillars. That's genuine impact. But her Time and Talent scores are mostly 2s and 3s. She attends one gala per year, serves on no boards, hasn't used her legal expertise for pro bono work. Her agency to act and create? Largely dormant.

When Maryanne saw her matrix for the first time, she said something I've heard from many high-earning donors: "I thought I was doing a lot. But I'm really just writing checks."

Her agentic philanthropy assistant, Phay.ai, surfaced an opportunity she hadn't considered: pro bono legal review of charter school contracts for an education nonprofit she already funds. Same cause. New contribution. The AI didn't replace her judgment; it expanded her field of vision. Maryanne galvanized around her agency to act.

She's now three weeks into that 90-day Agentic Island. Her projected score after completing it: 185. More importantly, she's using parts of herself that writing checks never touched.

David: The Time-Rich Contributor (Score: 142)

David retired three years ago from a career in logistics management. He volunteers twenty hours a week. Food bank sorting, Habitat builds, hospital visits, youth mentoring. He's everywhere.

Here's his matrix:

	TIME	TALENT	TREASURE	TRUST
EMPOWER	8	4	3	4
INNOVATE	6	3	2	3
COLLABORATE	7	4	3	3

David's Time scores are excellent. Organizations love him. But his Talent scores are moderate despite thirty years of logistics expertise. He's sorting cans at the food bank when he could be redesigning their entire distribution system. His agency to create isn't being used.

His Trust scores are the real missed opportunity. David knows dozens of other retirees with time and skills. He hasn't recruited a single one. His agency to connect is sitting idle.

When David saw his matrix, he was defensive. "I'm giving twenty hours a week! How is that not enough?"

It's not about enough. It's about alignment.

His agentic philanthropy tools helped him see the pattern: high effort, underleveraged expertise. They also helped him design a solution. His Agentic Island: create a volunteer logistics coordinator role for the food bank, then recruit and train three other retired professionals to fill similar roles at partner organizations. Moreover, there was no question that David tapped into his agency to create.

He's not just contributing anymore. He's multiplying. His agency to connect is finally activated.

GiveIQ Scoring™

Human Agency × Agentic Philanthropy

GiveIQ™

Human Agency

120 POINTS MAX

	Time	Talent	Treasure	Trust
Empower	1–10	1–10	1–10	1–10
Innovate	1–10	1–10	1–10	1–10
Collaborate	1–10	1–10	1–10	1–10

$$\times$$

Agentic Philanthropy Multiplier

1.0× to 3.0×

Multiplier	Description
1.0×	**Traditional philanthropy.** No agentic tools. You do the work yourself.
1.5×	**Basic AI assistance.** Using Phay.ai for recommendations, research, and matching.
2.0×	**Networked agency.** Phay.ai connects to vertical agents, surfaces opportunities you'd never find alone.
2.5×	**Cairn integration.** Your contributions build on collective intelligence; you draw from and add to the Global Charity Commons.
3.0×	**Full agentic ecosystem.** Multi-agent collaboration, verified impact loops, your efforts catalyzing systemic change.

120 × 3.0 = 360

Human Agency × Agentic Multiplier = Maximum GiveIQ Score

Keisha: The Young Connector (Score: 118)

Keisha is twenty-six, two years into her first real job as a development coordinator. She doesn't have much money to give. But she's passionate about getting communities of color outdoors.

Here's her matrix:

	TIME	TALENT	TREASURE	TRUST
EMPOWER	4	5	2	6
INNOVATE	3	6	2	5
COLLABORATE	4	5	2	7

Her Treasure scores are low. She gives what she can, about $50 a month. No shame in that.

But look at her Trust scores. She convinced her company to sponsor a kayaking event. She's brought five friends to volunteer days. She connected two nonprofits that are now running a joint campaign. Her agency to connect is her superpower.

Keisha sometimes feels like she's not a "real" philanthropist because she can't write big checks. Traditional philanthropy metrics would agree with her. GiveIQ doesn't.

Her Trust-Collaborate score of 7 is higher than Maryanne's or David's. She's a multiplier. And agentic philanthropy tools help her see that clearly, validating what she already sensed but couldn't prove.

Her growth path isn't about finding more money. It's about deepening what she's already good at and building her agency to act so she grows her own capacity over time.

Robert: The Plateau (Score: 168)

Robert is forty-three, a mid-level manager at a healthcare company. He's been involved in philanthropy since college. Serving on boards, giving consistently, volunteering regularly. His score has been between 160 and 175 for the past four years.

Here's his matrix:

	TIME	TALENT	TREASURE	TRUST
EMPOWER	5	4	5	4
INNOVATE	5	5	4	4
COLLABORATE	5	4	5	4

Robert's matrix is remarkably even. Mostly 4s and 5s across all twelve cells. He's doing a little of everything, nothing badly, nothing exceptionally.

And he's stuck.

Robert's plateau isn't about effort. He's working hard. It's about diffusion. He serves on three boards, gives to twelve organizations, volunteers for whatever's asked. He's spread so thin that none of his agency, to act, create, or connect, goes deep enough to matter.

When Robert saw his matrix, he immediately recognized the problem. "I'm a mile wide and an inch deep."

His breakthrough came from subtraction, not addition. With help from his agentic philanthropy tools, he identified where his contributions were actually creating impact versus where he was just showing up. He resigned from two boards to focus on one. He consolidated his giving from twelve organizations to four. He committed to a single Agentic Island: redesigning the patient experience program at a community health center.

Six months later, his score was 198. Not because he was doing more, but because he was doing less with greater depth. All three forms of agency, finally concentrated.

What These Four Stories Tell Us

The total score matters less than the pattern. And the pattern reveals where your agency is flowing and where it's blocked.

Maryanne had the resources but not the engagement. Her agency to act and create sat dormant while her treasure did all the work.

David had the time but not the leverage. His agency to connect sat idle while he sorted cans.

Keisha had the connections but not the confidence. Traditional metrics made her feel small when her real contribution was enormous.

Robert had everything but focus. All three forms of agency, diluted across too many commitments.

In each case, agentic philanthropy helped them see what they couldn't see alone. Not by replacing their judgment, but by amplifying their vision. That's the promise: AI that helps you become more of who you already are, not less.

Common Patterns You Might Recognize

The Checkbook Philanthropist: High Treasure scores, low everything else. Financial agency is flowing, but the agency to act and connect remains untapped. Like Maryanne before her awakening.

The Lone Wolf: High Time and Talent scores, low Trust. Dedicated volunteers who do excellent work alone. Like David before he started multiplying.

The Multiplier: High Trust scores across all pillars. These are the people who make ecosystems work. Like Keisha, often undervalued by traditional metrics.

The Dabbler: Moderate scores across everything, nothing distinctive. Like Robert before he learned to subtract.

How the Score Works

The mechanics are simple. Three pillars (Empower, Innovate, Collaborate) times four T's (Time, Talent, Treasure, Trust) equals twelve cells. Rate each cell 1-10 based on the past 90 days. These possible 120 human agency points are then multiplied by possible agentic philanthropy points. Maximum score: 360.

Your initial score is self-assessed. That's the starting point. As you complete Agentic Islands and your contributions get verified by organizations, documented through impact data, or validated by peers you've recruited, your score carries more weight. Verified contributions build trust in the system and in yourself.

The detailed scoring methodology is in the appendix if you want it. But the formula isn't the point. The pattern is the point. Where is your agency flowing? Where is it blocked? What would it look like to open a new channel?

Trust in the Age of AI

Of the four T's, Trust deserves special attention.

In the AI era, trust is the health of our relationships. With ourselves, our communities, and the causes we serve. Anyone can puff up their social media presence. We've all seen people showcasing restaurants and vacations while struggling in their personal lives. That works in the short term. But there's always a moment of truth.

The GiveIQ Trust principle is different. It's not a credit score or a branding metric. It's a measure of how you're strengthening the connective tissue of philanthropy itself. When you recruit a friend, connect two organizations, or share what you've learned along your Purpose Path, you're adding to the Global Charity Commons. You're improving the trail for those hiking after you.

Yes, AI can help build online presence. It already does. But agentic philanthropy is about something deeper: AI that helps you build real trust through real contribution, then amplifies that trust across networks you couldn't reach alone.

Your Score Is a Compass

I want to be clear about something: Your GiveIQ score is not your worth as a human being. It's not a competition. It's not a judgment.

It's a compass.

A compass doesn't tell you where you should go. It tells you where you are and helps you navigate toward where you want to be.

A score of 95 isn't bad. It's a starting point. A score of 280 isn't perfect. It's a snapshot of someone at a particular moment on their purpose path.

What matters is direction. Are you growing? Are you moving toward alignment between what you say you care about and how you actually contribute?

The agency to act. The agency to create. The agency to connect.

Where is yours flowing? Where is it ready to be unlocked?

That's what the score helps you see. And that's what agentic philanthropy helps you do something about.

EIC: The Engine of Value Creation

Culture is the mindset. EIC is the method.

EIC VALUE FLYWHEEL

Empower
the agency to **Act.**

Innovate
the agency to **Create.**

Collaborate
the agency to **Connect.**

- Human capacity
- Organizational adaptability
- Collective intelligence

Empower develops human potential.

Innovate transforms capacity into new possibilities.

Collaborate amplifies and scales outcomes.

Staying Human in the Digital World

This is what I mean by "being smarter" in the age of AI. It's not about having more information or faster processing. It's about maintaining the discernment to know what deserves your engagement. It's about keeping your critical thinking sharp when algorithms want to think for you. It's about staying present to real human need when virtual worlds are engineered to capture your attention.

AI can help you discover causes that align with your values. It can match your skills to organizations that need them. It can optimize the timing of your contributions. It can connect you with collaborators you'd never have found on your own.

But AI cannot replace your actual decision to give. It cannot replace your hours spent volunteering. It cannot replace the expertise you bring or the trust you build. It cannot replace your presence at the board meeting, your voice advocating for change, your hands doing the work. And importantly, it cannot replace your real relationships with friends and humankind.

Your human agency, your active, engaged, conscious choice, is what creates impact. AI can amplify it, but it cannot substitute for it.

Human Agency x Agentic Philanthropy = Societal Impact

That formula only works if the first term remains strong. If we abdicate our agency, if we let AI decide what we care about, where we contribute, how we engage, then we've created feudal AI, not agentic AI. We've become digital serfs on someone else's estate.

But if we actively employ our human judgment, use AI as an amplifier rather than a replacement, and ground our engagement in real-world action that we can measure and improve, then we create something transformative.

You don't have to summit Mount Washington on your first day. You make sure you are prepared and have learned some basics before you venture out. You can start with a day hike, complete a meaningful project, and build from there. But you do have to actually walk/hike the trail. You do have to show up in the real world. You do have to let your humanity shine through.

Keeping it real, staying actively engaged in the physical world of actual human need, that's the essence of remaining human in a digitally overloaded world.

The trail system exists. The maps are drawn. AI can help you navigate.

The only question left is: Are you willing to do the hike?

When Your Main Trail Gets Blocked

Six weeks into her digital equity Agentic Island, Gus sat with her mentor reviewing progress. The work was going well. But something was nagging at her.

"My dad got laid off last week," she told her mentor. "He's been in customer service for twenty years. They replaced his entire department with an agentic AI system." They had been replaced like tens of thousands before them by *Virtualogie,* a new agentic AI customer service platform that had been taking the world by storm.

That evening, Gus found herself thinking about Purpose Paths differently. It was one thing for a student with options to choose a trail. But what about someone like her dad, whose trail had just been bulldozed?

She knew he'd spent twenty years answering phones, resolving complaints, navigating corporate systems on behalf of frustrated customers. What she didn't know was whether that experience added up to anything beyond "customer service rep" on a resume.

So she asked him. "Dad, outside of work, what have you actually done the last ten years?"

He looked at her like it was a trick question. "What do you mean, done? I worked. I paid the bills."

"No, I mean, little league coaching, the neighbors you helped with insurance stuff, the new people at work you always mentored."

He started listing things, reluctantly at first. Eight years coaching little league. Three neighbors he'd walked through insurance claims nightmares. Mentoring new customer service reps even when his managers didn't ask. Translating medical bills for his elderly mother. Helping his brother-in-law navigate disability paperwork.

"That's not work," he said. "That's just... stuff I did."

"That's your trail record," Gus said. "You've been walking a purpose path for years. You just never thought of it as a path."

When he finally mapped those trails using the GiveIQ approach, a different picture emerged. He cared about helping people who felt overwhelmed by bureaucracy. He'd been choosing to do this work for free, repeatedly, for years. He had twenty years of proof he could translate complex systems, stay patient with confusion, and follow through until problems were solved.

Three weeks later, Gus's dad agreed to try something small. A 90-day Agentic Island. A local nonprofit supporting immigrant entrepreneurs needed someone to help five business owners navigate small business licensing and insurance requirements. Volunteer work, unpaid, three hours a week.

Everything in him resisted. "I should be applying for jobs, not volunteering."

But Gus pushed. "You'll spend ninety days either sitting at home getting more discouraged, or you'll spend ninety days proving you can do meaningful work outside that customer service department."

He started the island. And within two weeks, something shifted. By week six, one of the business owners asked if he'd consult for them part-time, paid. Another connected him to a community college that needed someone to teach a course on navigating government services.

None of these came from job applications. They came from demonstrated capacity. From his trail record becoming visible.

The Age of Purpose: Reclaiming Community

Over the past several decades, these community caring values have eroded. Robert Putnam documented this decline in *Bowling Alone*[^9]—the collapse of civic associations, the emptying of community halls, the retreat into private life. Technology accelerated the shift. Social media platforms create illusions of connection while diminishing real relationships. We scroll past thousands of faces but know fewer neighbors than our grandparents did.

The consequences are measurable: rising loneliness, deepening polarization, crumbling trust. This isn't coincidental, it's the direct result of trading civic engagement for personal isolation. We've built an economy of attention that profits from our disconnection.

But here's what I've learned from thirty years in nonprofits: the impulse to help hasn't disappeared. It's been suppressed, misdirected, waiting for new channels. When people find meaningful ways to contribute, when they discover their Purpose Path, that dormant civic spirit roars back to life.

Charity Autonomy is designed to reawaken it.

CHARITY AUTONOMY

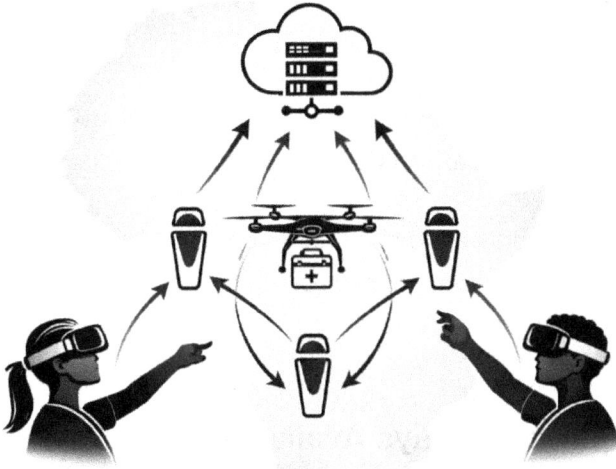

🔑 Key Takeaways

☑ **Charity Autonomy democratizes purpose.** With AI as a helpmate (not a decision-maker), individuals can directly contribute their time, talent, treasure and trust without waiting for permission or institutional approval.

☑ **The job market crisis is a purpose opportunity.** As AI displaces entry-level positions across once-secure fields like accounting, law, and computer science, universities must pivot from transactional degrees to lifetime Purpose Paths.

☑ **From transactions to transformation.** Charity Autonomy replaces one-time donations and clicktivism with sustained engagement through tools like Agentic Islands and personalized Purpose Paths. Like Commonwealth Corps proved in 2006, lowering barriers unlocks latent capacity, and with today's AI tools, we can scale that model globally, turning every contributor into a skilled participant whose competence compounds with each project.

Understanding Charity Autonomy

Charity Autonomy is the power to do good directly, freely, meaningfully. No board approval. No mandatory gala. Just work that matters.

This isn't rebellion, it's progression in nonprofit design. With today's AI tools, you can track impact, share progress, and connect to others. Charity Autonomy says: You need not wait. Begin now.

The AI Helpmate: Always Available, Never in Charge

The ambient nature of artificial intelligence will mean we have a helpmate available 24 hours a day, 7 days a week. Not a boss. Not a decision-maker. A helpmate, ready when we need guidance, silent when we don't. Moreover, these ambient AI agents and agentic philanthropy will also watch one's blindside, including identifying and reducing risk.

Your Charity Autonomy uses those AI and other tools to make your contributions of time, talent, treasure, and trust invaluable to the causes you care about. AI can help you find the right organization. It can match your skills to urgent needs. It can track your impact over time. It can connect you with collaborators you'd never have found alone. But the decision to act, the commitment to show up, the meaning you derive from the work, that remains entirely yours.

From One-Time Clicks to Meaningful Impact

Consider our responses to extreme weather events and disasters around the world. A hurricane devastates a coastal community. An earthquake levels a city. Wildfires consume entire regions.

The typical response? A one-time credit card donation prompted by a news cycle. Click, contribute, forget. The disaster fades from headlines, and so does our engagement.

Charity Autonomy engages us and our human promise at another level, from not doing something, or making a one-time contribution, to one of meaningful, sustained mission impact. The power is now in your hands, perhaps to do something that maybe a small nonprofit would have been able to do before.

What if, instead of a single donation, you joined an Agentic Island focused on disaster preparedness in vulnerable communities? What if your professional skills, construction, logistics, counseling, communications, were matched to organizations that need exactly what you offer? What if you could see how your 90-day commitment connected to a larger network of contributors, each bringing different capacities, testing scenarios, all coordinated toward resilience?

That's the difference between transactional charity and Charity Autonomy. One makes you feel good for a moment. The other changes who you are.

The Three Core Values of Charity Autonomy:

As we enter a world of abundance, with increased discretionary time and improved digital infrastructure, philanthropy needs to stop being exclusive.

Charity Autonomy unlocks three things:

- ❑ Purpose over prestige
- ❑ Contribution over credentials
- ❑ Collaboration over control

The Arc of Charity Autonomy:

- ❑ **Past:** Gatekeepers and elite boards controlled access and direction.
- ❑ **Present:** Platform tools emerge. Donor-advised funds, crowdfunding, online organizing.
- ❑ **Next:** AI-driven personal giving agents and curated purpose platforms.

❏ **Future:** A World Charity Model. Fully distributed global charity autonomy with cross-border collaborations.

AI and Nonprofits: Enabling the Purpose Path

Operational Challenges in the Nonprofit Sector

This is a microcosm of the broader problem. Nonprofit organizations are battling to raise so much money each year while working in parallel to achieve their mission impact. Consequently, they usually do not have the white space or resources to re-design the way things work. The culture is usually a heads-down approach toward achieving annual goals. The common annual strategy is for nonprofits to plow ahead in the most linear way and go with the tried-and-true wealthy people who have been dedicated and loyal donors, in some cases for years. Then, these charities follow up by asking those donors to help broaden the circle of giving. Asking "who else should we invite to participate?" This approach doesn't really open things up to all of us, does it? It simply offers conservative concentric rings of connection to the current donor.

This, in turn, weakens the idea generation and diversity of folks contributing, putting a chokehold on strategy. There are no limits on ideas; we have stockpiles of ideas. Leaders will say to me that it is unrealistic to develop, test, and enact new ideas like start-ups do. Clearly, every entity, for-profit and nonprofit, has constraints on which ideas it chooses to advance. But by using agentic philanthropy to help reform the way causes and mission-driven organizations work, we can accelerate the introduction and iteration of good strategies while shifting to a democratization of philanthropy.

Charity Autonomy is an era in which everyone can have a stake in driving social good and taking the reins of powerful creation through AI. Not just the privileged few but legions of curious, committed, smart, aligned, intuitive, and action-oriented servant leaders. I love the idea of each of us stepping up our individual active meaning and purpose. AI then fuels this Purpose Path within the GiveIQ ecosystem, with the tools of giving, organizing, learning, testing, iterating, collaborating, and leading. The GiveIQ vision imagines a world where purpose is everyone's priority, a new ecosystem where individuals from all walks of life can plug into their values using AI as a helpmate.

I liken it to switching between driving a Tesla from driving a traditional car. In most vehicles, the right-column of the steering wheel features a stick/control that operates the windshield wipers, but in a Tesla, that same stalk is the gear selector. Acting on muscle memory in the rain could inadvertently shift into drive at 55 mph. Habits built for one system don't always translate safely to another. The same principle applies when working with AI agents.

Initially, managers may delegate to multiple agents per day, giving precise, no-wiggle-room instructions. Over time, though, agents may learn preferences and begin to anticipate needs. Progressive agentic AI improvements will be made through reinforcement learning, autonomous learning and eventually recursive self-improvement (when AI can improve its own architecture). And our trust and interface with agents will increasingly be on par with the trust or interfaces with other humans.

The future that GiveIQ imagines is one where AI strengthens our ability to act with purpose. Every nonprofit, foundation, and mission-driven organization has an opportunity to design experiences that align with the Purpose Path of their donors, volunteers, employees, and constituents. Rather than focusing solely on transactional engagement, charities can use AI to offer personalized Purpose Paths that keep people engaged in their mission over time.

With agentic philanthropy, donor engagement becomes proactive, iterative in real-time, and more aligned with individual personal motivations. AI-powered models will create customized pathways for each person, helping an organizational novice get started or the expert challenge themselves. And while agentic philanthropy supports a better volunteer or donor experience, the same will be said for how agentic philanthropy empowers employees. Customized pathways sketch out how staff's concrete objectives align with a strategic plan and greater mission impact.

Phay.ai as a Purpose Navigator

Unlike traditional AI recommendation engines, an AI agent like the hypothetical Phay.ai isn't a deterministic system but a dynamic navigator that refines a user's Purpose Path through continuous learning, iterative feedback, and adaptive modeling.

Phay.ai might be comparable to a personal museum curator. That curator knows your likes and dislikes and what you'd like to experience. Great curators are great listeners too, personalizing the experience. In touring the Louvre for instance, let's say a curator or docent named Lara asks us question along the way or maybe gamifies the experience asking kids to look for clues in important pieces of art.

Navigating purpose will grow as a chief way of personal growth and learning, including for youth. My nine-year-old daughter and others her age use logarithm-driven educational programs like DreamBox and AdaptedMind, to identify where and how to improve learning.[8] Platforms like these employ adaptive learning algorithms that function as real-time tutors. For charities, similar tools may offer relevant and tailored opportunities, while building individual and organizational capabilities. And digging deeper, this ensures that participation remains personally fulfilling and impactful.

What's Your GiveIQ?

For generations, charity work was largely the domain of the privileged. Wealth opened the door to boardrooms, naming rights, and influence over direction. It has been a pay-to-play environment for so many charities. As discussed, this was a fact of life for so many nonprofits – they traded board seats and access to governance for organization-sustaining funding. Now, a new chapter is possible.

We're standing at the dawn of Charity Autonomy: an era where individuals are empowered to do as much as they want and to do it well. Powered by AI and digital tools, everyone can participate meaningfully in philanthropic work. This time is one

---

8   DreamBox Learning, "Intelligent Adaptive Learning for K-8 Math," Bellevue, WA: DreamBox Learning, 2024. https://www.dreambox.com

that truly amplifies what one person can do while tapping into the collective intelligence and combinatorial power of a network of folks with similar cause interests and visions. Consequently, this new block of powerful philanthropic contributors disassembles the ancient philanthropic cartel, long the domain of the wealthy and well connected.

Commonwealth Corps: Charity Autonomy Before We Had the Name

Long before I developed the GiveIQ approach, I saw what happens when you lower barriers to service.

In 2006, I attended an event and had a chance to meet Deval Patrick and his wife Diane. Deval was running for Governor of Massachusetts. We had a great chat on ways to boost volunteerism in the state. At the time, I was serving as Chair of the Massachusetts Service Alliance, the state's Commission on Service and Volunteerism. Each state had a similar organization that led service and volunteerism efforts, principally distributing AmeriCorps and VISTA programs.

I pitched a new idea to the Patricks: Commonwealth Corps. We wanted to create something that provided more access to folks who wanted to serve in their communities. The national program usually necessitates at least a whole year commitment and a move to a new state, along with housing and other logistics. Commonwealth Corps would allow folks to sign up for shorter stints and even identify the nonprofits they'd like to work with locally. It could be part-time and focused on a number of the state's most pressing needs.

The Patricks liked the idea and asked terrific questions. But I thought the conversation was the end of that topic, I'd need to reintroduce it in the new year to whomever won the governorship.

Deval Patrick did win. And in his first week in office, one of his team members called me. She said, "The Governor wants to do that program you talked with him about."

I was thrilled, to say the least.

Over the next few months, we worked with great champions for service in the legislature, and that June we passed the Commonwealth Corps bill. Since then, hundreds of people have been working locally, doing great work to strengthen Massachusetts communities.

What made Commonwealth Corps work? The same principles that underpin Charity Autonomy today:

- ❑ **Lower barriers:** You didn't need to uproot your life for a year. You could serve part-time, locally, on your own terms.
- ❑ **Match passion to place:** Contributors could identify the nonprofits they wanted to work with, not just accept an assignment.
- ❑ **Bounded commitment:** Shorter stints meant more people could participate, students, working parents, retirees, anyone with something to give but not a year to give it.
- ❑ **Local impact:** The work strengthened actual communities where contributors lived, creating visible, tangible results.

That was 2006, before smartphones were ubiquitous, before AI could match skills to needs, before digital platforms could coordinate thousands of contributors in real time. Imagine what's possible now. It was right around the first time I was thinking about the idea for intelligent philanthropy and subsequently GiveIQ.

Charity autonomy is the basic cornerstone of the GiveIQ vision. Charity Autonomy means a mother in Malawi, a teen in East Boston, or a retiree in County Cork, Ireland can act on their GiveIQ. It means people can pool resources, launch charity campaigns, start impact projects, or advise others, all with personalized AI assistants that help them navigate the complexities of giving.

Like any learned skill, competence in charity builds confidence. As we improve and align our efforts, we achieve more, both individually and collectively. Like our own personal flywheel, we become more competent and valuable to the next cause or project we undertake.

The Ellen Effect

Commonwealth Corps proved that lowering barriers to service unlocked latent capacity. But there's another dimension to Charity Autonomy that's becoming urgent: what happens when the traditional career ladder disappears?

The headquarters attic space for Habitat for Humanity at the Covenant Congregational Church usually surprised people. "This is your office?" folks would ask incredulously. I loved it though. We were a couple blocks away from the Forest Hills Orange Line train into downtown Boston. It gave me a chance to meet with prospective donors in Boston's financial district or at a local coffee shop. The only difficulty about my location out at Forest Hills was getting office volunteers. We had some folks that would stop in regularly and help us stuff envelopes for a mailing or come to an evening meeting to mentor new Habitat homeowners. But this was considered just enough out of the way to preclude many of our suburban volunteers.

We tried regularly to send out volunteer requests in our emails. I would collect business cards at functions and other events and then enter the email addresses from those cards when I got back to the office. This was my nascent database. And Constant Contact was our early CRM. One day I received an email from a woman named Ellen in reply to our request for a volunteer grant writer. Ellen said she was a good writer, but had never written a grant. She was willing to learn though and wanted to give grant writing a shot. I needed help, so I agreed to meet her.

Sure enough, she was a great writer. We'd work together on grants that were usually in one of three simple themes. Perhaps it was a grant for funds to acquire land for new Habitat homes. Another to help us with day-to-day operations, including getting some basic computers or construction site tools. The most common were requests to help us build houses. Construction materials, funds for subcontractors like electricians or plumbers, appliances and more.

Ellen became so adept at writing grants that she came in one day and asked if I would write her a recommendation for a job she had applied to at a university. I agreed – but she really didn't need my help. Based on her grant writing, that university agreed to give her a trial run. And as they say, the rest is history. She has gone on to an amazing career in the development field.

In today's job market, I think volunteering like Ellen did should be a requirement for college students. Their GiveIQ might be entry level, but after working alongside great organizations and for great causes, I can only imagine the skills and confidence they will build. As entry-level jobs get scarce and more competitive, experiential learning is a must in this new AI era. From college students to older workers, you can gain valuable experience to help get you in the door somewhere else or start your own gig – fully deploying your Charity Autonomy.

The hard numbers on how AI is affecting entry-level jobs are difficult to quantify. Some say the tightness isn't from AI taking those jobs directly, but from companies redirecting resources from internships, co-ops, and first-time positions to the purchase of new AI and other technologies. I do believe it won't take long for AI agents to outcompete recent college graduates. This should produce fresh worry among universities.

I wanted to hear how companies are dealing with this. At Dell, for instance, employees told me their department mandated they use AI twenty-five times a week to assist them with their work. These employees aren't encouraged to use AI - they're mandated. This seems rudimentary, but it's not a bad tactic to get folks to throw themselves into the deep end. One new employee talked about how "prompt wrangling" had become a daily thing.

There are ominous rumblings from computer science graduates from some of the nation's top programs. Coding was once a protected space for them and a guaranteed good living. It seems like there's a daily encroachment on computer coding though. Vibe coding is just the beginning. Our natural language interfaces get better each day. Novices can create websites on Replit and Lovable now with simple prompting. And soon we will all be able to create the kind of apps one could only dream about just a few years ago – just by talking with a next-generation Alexa or Siri. GiveIQ's Phay.ai for instance might be the Charity Autonomy helpmate of choice, asking us how we want to add value to a charity's mission or contribute to a cause.

Ellen didn't know it at the time, but she was building her career insurance in that attic office. The question now is whether we can create that pathway intentionally - at scale - before the job market forces the issue.

The Purpose Year

The Talent Purpose Path Pipeline

With more time on our hands there will be a bigger supply of volunteers. This volunteer workforce will consequently need to do more to distinguish themselves, including working to be identified as top talent. This will be a good thing for nonprofits that embrace GiveIQ and exponential impact goals.

If experts and entrepreneurs like Meta's Mark Zuckerberg think we are close to superintelligence, that means we are also close to a superintelligence-empowered volunteer. Zuckerberg said "We envision deeply personalized super-intelligent assistants, AI companions that understand you, know your goals, and help you become the person you aspire to be. These are not tools for automating tasks, but partners empowering your creativity, connection, and personal growth." Meta, Google's Gemini and Grok are designing new waves of superintelligence interfaces. It will be essential for the smartest and most helpful volunteers to be fully up to speed on the latest and greatest intelligence tools and continuously learning, to be amongst the most "talented."

Universities are facing a reckoning they didn't anticipate.

This will be one of the defining realities of tomorrow's graduations: a slow and steady decline in those careers we thought were bulletproof – law, accounting, marketing, and computer science. The instinct for many colleges will be to move incrementally along. But this is not the time for glacial change. How do they promote a lifetime of purpose? University leadership can't afford to be oblique. Things are changing and higher education needs to change too.

Let's say you are a recent accounting major graduate and have had no luck getting an entry-level role at local accounting firms – never mind the big three firms. The jobs you thought were there are now being done by AI accounting agents, who are managed by a superior AI agent, who in turn has a human boss. I know some of you are in disbelief, but this scenario is coming fast. So why not consider how you can use those accounting (or insert any other major here) skills in the field of purpose? What if universities set their sights on making the junior or senior year a year of Charity Autonomy - inspiring young people to spend a year working for a nonprofit and honing their Charity Autonomy? I know it sounds idealistic - and I've painted an extreme scenario - but what if Charity Autonomy was the only way to make a living for a significant segment of graduates?

What an exciting opportunity for alumni to fund and participate in this program. Through the GiveIQ ecosystem - Agentic Islands, Web 3.0, and the Global Charity Commons - you not only have alumni donating, but you have them actively engaged in good work around the world. Imagine what happens to your university's brand when a hundred alums in branded sweatshirts get off a plane after spending two weeks working with students in a village, building a new clean water system. And this is what we are talking about when we mention qualia. Wow – right?

Let's move higher education from a four-year degree and you're done - to a lifetime subscription model of learning and contribution. The university's responsibility isn't just to educate people, it's to give them some of the tools they need to realize their full potential and a meaningful life. With that potential in mind, coupled with the power of AI, each university needs to ask themselves how they will set their graduates on their respective Purpose Paths. Curriculum that strengthens their students in EIC principles. Agentic Islands that give them platforms to act. And a Purpose Path that doesn't end at commencement - it begins there.

Ask Yourself

For Individuals:

1. What's your GiveIQ? And with it, what problems will you solve?

2. Who will you lift up? How will you collaborate?

3. What role does AI currently play in my decision-making, and how can I ensure it enhances rather than dictates my purpose?

4. What small, consistent actions can I take today to build a more purpose-driven life?

For Nonprofit Leaders:

1. How can charities use AI to deepen human connection rather than automate engagement?

2. Are we designing our programs and donor interactions with purpose-driven personalization?

3. How can we foster a culture of agency, empowering individuals to take ownership of their charitable work?

AGENTIC ISLAND

CLOUD COMPUTING · AI AGENTS · BLOCKCHAIN · ITERATION

AGENTIC ISLAND: NEW NONPROFIT TEAMS

🔑 **Key Takeaways**

- ☑ **The Kitchen Island Metaphor:** Like a bustling kitchen island where chefs gather around shared ingredients, Agentic Islands bring humans and AI together in focused, collaborative spaces.

- ☑ **AI as Sous Chef, Human as Executive Chef:** AI handles coordination and repetitive tasks while humans provide vision, empathy, and mission alignment.

- ☑ **Fluid Team Design:** Move from rigid hierarchies to modular teams that form around projects, dissolve when complete, and reform as new needs emerge.

Julia Child and the Legacy of the Kitchen Island

Few figures are as synonymous with the art of the kitchen island as the legendary American chef Julia Child. Towering at 6'2" with a magnetic presence that transformed cooking into a joyous performance, Julia Child captivated audiences on her long-running public broadcasting show. Living just a few miles from me in the Boston area for forty years, she shared invaluable tips—"don't crowd things in the pan," "cook with copper," "don't take yourself too seriously," and "keep your kitchen knives sharp." These insights not only refined the craft of cooking but also turned the kitchen island into a vibrant hub of creativity and learning.

In the spirit of Julia Child's legacy, the Agentic Island becomes a dynamic meeting place for experts and enthusiasts alike, a space where knowledge is shared generously and innovation is born from collaboration. On this metaphorical island, specialists from diverse fields come together, much like chefs in a bustling kitchen. For instance, an expert in diabetes fundraising might join forces with medical researchers, data scientists, and case study authors. Together, they combine world-class research, refined data drawn from vast data lakes, and cutting-edge algorithms, or even craft personalized algorithms tailored to individual giving patterns in ways never seen before.

Just as Julia Child invited her viewers into her kitchen to learn and experiment, the experts gathered around this island each contribute a unique piece to the puzzle. Their collective expertise creates compelling products or services that redefine the philanthropic landscape. The legacy of Julia Child serves as a powerful reminder that the Agentic Island is not just a physical space; it is a state of mind. It is about embracing creativity, fostering a spirit of generosity, and recognizing that every contribution is vital. Just as Julia's practical tips and engaging personality made cooking accessible and enjoyable, the Agentic Island metaphor makes the integration of AI and human ingenuity transformative for nonprofits.

Practical Implementation: Bringing the Agentic Island to Life

Turning this metaphor into a practical reality means integrating a thoughtful suite of digital tools and strategies that support both data-driven decision-making and genuine human connection. It starts with developing a centralized platform that seamlessly blends the digital dashboard with collaborative tools like real-time chat, project boards, and AI-powered insights. This platform becomes the nonprofit's central hub, a place where all activities converge and collaboration flourishes.

To keep the momentum going, nonprofits can implement AI check-ins, where intelligent agents proactively engage with team members. These agents offer updates on key metrics, suggest evolving priorities, and provide insights that guide day-to-day activities, acting almost like digital collaborators. Next, organizations should encourage daily scrums and opportunities for teams to "meet at the island," whether virtually or in person. These touchpoints create a rhythm of quick reviews and ground everyone in a shared sense of progress and purpose.

By keeping channels open across geographic and organizational boundaries, nonprofits can foster a culture where every voice is heard and every idea has a place. This creates a deeper sense of belonging and shared ownership. And most importantly, organizations should always be ready to learn from each recipe. With AI helping to analyze what worked and what didn't, teams can iterate and improve. Insights can reveal which contributors, or "key ingredients," were essential and should be present at the Agentic Island for future initiatives.

Augustina Jones as Executive Chef

Gus Jones's company, Cyber Guardians, devoted serious time to social impact, both to support corporate responsibility goals and to create project-based experiential learning. As Gus weighed in on which causes the company should volunteer to support, the Southeast Asian Nations Home Care Network (ASEANCare) quickly rose to the top of the list.

ASEANCare supports tens of thousands of households caring for loved ones at home. In one striking example from the Philippines, a family transformed a small downstairs dining area into a cozy bedroom for their 95-year-old matriarch, Edna.

ASEANCare had created a seamless balance between full-time ground staff, volunteers, humanoid robots, and drone-based medical delivery services. One of the most valuable tools underpinning this orchestration was their adoption of the Agentic Island concept, a metaphor that brings to life how diverse teams and technologies can converge in one virtual space to generate project synergies.

Gus and her team established the ASEANCare Agentic Island by first analyzing key daily tasks, defining core objectives, and understanding both customer needs and how staff and volunteers interacted with the organization.

A "team hub" was created. A good metaphor for the team hub is the kitchen island: a bustling kitchen island brings chefs together around a shared pantry of tools and ingredients. The team hub model enabled teams to collaborate in one virtual space, sharing ideas and resources in a fluid, dynamic manner.

In this framework, the team hub lead is the human in charge who orchestrates the collaboration between people and agentic philanthropy. For the roll-out of ASEANCare's new paramedic drone program, Gus Jones served as that hub lead, like an executive chef might do in the kitchen. Gus and the Cyber Guardians were happy with their work and the subsequent bump up in their GiveIQ scores. The most satisfying element of the experience was improved care for very low-income senior citizens in Southeast Asia.

The Agentic Island Metaphor

In today's fast-evolving landscape, the agentic world of AI is not a distant promise; it is here and now. These AI agents are capable of reflection, planning, and iteration. They serve as collaborators, forming a dynamic ecosystem of multi-agent systems that write code together, solve mysteries, and adapt to tasks at speeds that frequently outpace human capabilities.

Within the nonprofit and philanthropic sectors, the implications matter. Recent studies reveal that more individuals are already leveraging AI in their daily tasks than are provided access by their organizations. Waiting for the "perfect moment" to integrate AI is no longer an option.

The Kitchen Island: A Space for Connection

In many homes there is a kitchen island. It's the place where we organize to cook a good meal. Within reach are mixing bowls and spices, special ingredients and measuring cups. It's also a place where we may solicit help and advice—"Does this look done to you?"

The kitchen island is a place of sharing and teaching. I think of the older Italian woman making homemade gnocchi, with puffing flour clouds as she kneads the dough.

GiveIQ imagines Agentic Island as both a physical and virtual space designed to foster collaboration, creativity, and purpose. Here, human agency and agentic philanthropy come together to collaborate around project-based objectives. It's more than just a digital platform; it's a shared environment where human and AI IQ converge to do meaningful work.

A defining feature is its **magnetic pull** — a constant, stabilizing force that keeps everyone connected to the organization's mission through:

- ❑ **Shared Accountability:** Transparent visibility into tasks and priorities
- ❑ **Creativity & Inspiration:** The space sparks spontaneous ideas and innovation
- ❑ **Mission Connection:** Keeping all contributors connected regardless of location

How People and AI Collaborate

In the Agentic Island model, **AI acts as a sous chef** — a capable and specialized assistant who supports human decision-making. These AI agents proactively surface insights, recommend next steps based on data, and handle repetitive tasks. They keep one's EIC at the fore, helping human participants tap into their time, talent, treasure and trust - to realize their promise.

Meanwhile, **humans take on the role of executive chefs.** They bring empathy, vision, and deep understanding of mission-context. The executive chef steers the ship toward the project goal while ensuring the journey reflects shared values. I think of one's cognitive skills, empathy and life-long learning being fully engaged around this role. There's also the opportunity for joyful connections with other humans participating in a shared objective.

The 4 T's Meet Agentic Philanthropy

Here's what makes the Agentic Island different from traditional volunteer coordination or team management: it's where human contributions, the 4 T's, meet AI amplification.

You bring: - Time: Your hours, your attention, your presence **- Talent:** Your skills, expertise, and professional knowledge **- Treasure:** Your financial contributions, however large or small **- Trust:** Your relationships, networks, and credibility.

Agentic AI amplifies: - Your Time by handling scheduling, coordination, reminders, and logistics, so your hours go further **- Your Talent** by matching your specific skills to where they're needed most, surfacing relevant resources, and helping you work at the top of your capability **- Your Treasure** by optimizing allocation, tracking impact, and connecting your giving to measurable outcomes **- Your Trust** by mapping networks, identifying connection opportunities, and helping you see how your relationships compound across the commons

The magic happens at the intersection. A retired accountant brings 10 hours a week and deep financial expertise. Without AI support, she might spend half that time on email coordination and figuring out what needs doing. With agentic philanthropy, she arrives at the island and immediately sees: here's the grant budget that needs review, here's the context you need, here are the three questions the development director is stuck on. Her 10 hours become 10 hours of actual high-value work.

A young professional brings limited treasure but high trust, she knows dozens of people in her industry. The AI doesn't just store her contact list; it identifies which of her connections might care about the current campaign, suggests personalized outreach approaches, and tracks which introductions led to engagement. Perhaps this leads to the agentic philanthropic AI designing a LinkedIn strategy or identifying for-profit partners who share the same corporate social responsibility goals. Her network becomes a strategic asset, not just a list of names.

This is the core promise of the Agentic Island: **human agency plus agentic philanthropy equals amplified impact.** You don't give up control. You don't become a cog in an AI-driven machine. You bring what only humans can bring, purpose, judgment, relationships, care, and AI handles the friction that usually dilutes your contribution.

AGENTIC
ISLAND
CHECKLIST

4 T's

☑ CHARITY
☑ AUTONOMY
☑ DIGITAL
TWIN

From Metaphor to Practice: Building Your Agentic Island

Turning this metaphor into a practical reality means integrating a thoughtful suite of digital tools and strategies that support both data-driven decision-making and genuine human connection.

It starts with developing a centralized platform that seamlessly blends the Dashboard with collaborative tools like real-time chat, project boards, and AI-powered insights. This platform becomes the nonprofit's central hub, a place where all activities converge and collaboration flourishes.

To keep the momentum going, nonprofits can **implement AI check-ins**, where intelligent agents proactively engage with team members. These agents offer updates on key metrics, suggest evolving priorities, and provide insights that guide day-to-day activities, acting almost like digital collaborators.

Next, organizations should **encourage daily scrums** and opportunities for teams to "meet at the island," whether virtually or in person. These touchpoints create a rhythm of quick reviews and ground everyone in a shared sense of progress and purpose. And there are new tech iterations in today's world of Microsoft Co-Pilot and other office assistants, that will manage and prioritize the daily work of these scrums.

By keeping channels open across geographic and organizational boundaries, nonprofits can foster a culture where every voice is heard and every idea has a place. This creates a deeper sense of belonging and shared ownership. Buying into a charity's vision is the first step to connect to a culture, but fostering the strength of this culture comes from frequent drops of connecting points – including personalized knowledge, wisdom and clarity.

And most importantly, organizations should always be **ready to learn from each recipe**. With AI helping to analyze what worked and what didn't, teams can iterate and improve. Insights can reveal which contributors, which "key ingredients"—were essential and should be present at the Agentic Island for future initiatives.

The Morning Ritual: Starting Each Day at the Island

As management expert Amy Webb has noted, "Embracing dynamic, adaptive systems is no longer optional; it's essential for survival in today's rapidly evolving landscape."

What if every nonprofit team started their day at the Agentic Island? Not a mandatory meeting, but a ritual check-in, five minutes to see the dashboard, review overnight AI insights, surface any blockers, and align on the day's priorities. Digging deeper, what if a curated philanthropic advisor like Phay.ai checked in with you – like your exercise or other daily goals – and presented some options for you to contribute time, talent, treasure and trust?

This isn't micromanagement. It's the opposite. When everyone can see what everyone else is working on, when AI handles the monitoring and flagging, when the system surfaces what needs attention, you don't need endless status meetings. You don't need managers chasing updates. The island creates shared awareness automatically.

The morning ritual becomes a moment of connection. You see colleagues you might not otherwise interact with. You catch a problem before it escalates. You offer help on something you didn't know needed help. The island creates serendipity by design.

Nonprofit Team Design for the AI Era

Traditional nonprofit structures often follow predictable patterns: hierarchical org charts, siloed departments, annual planning cycles that can't keep pace with rapidly changing conditions. The Agentic Island model suggests something different, fluid teams that form around specific projects, dissolve when complete, and reform in new configurations as new needs emerge.

This isn't about eliminating structure. It's about designing structures that can flex and adapt while maintaining mission alignment. It can be truly combinatorial too. Akin to a GANNT chart that's used to keep phases of a construction project on schedule, assignments build upon one another. The kitchen island doesn't eliminate the rest of the kitchen, it creates a gathering point that makes the whole system work more effectively.

Key principles for nonprofit team design:

Modularity: Build teams that can plug together in different configurations depending on the project at hand. And strengthen these teams by using the best agentic philanthropy coordination tools.

Transparency: When everyone can see what everyone else is working on, coordination happens naturally rather than through endless meetings. While AI will increasingly grow more ambient – working in the background – it would behoove nonprofits to design communications updates that are timely but not overwhelming. You want people kept in the loop, but only with relevant updates.

Bounded Autonomy: Give teams clear objectives and constraints, then let them figure out how to achieve the goals within those boundaries. Here, let me reinforce that humans are in charge. For instance, we mustn't let AI decide the pace to achieve these goals. What's a realistic human pace and how might agentic philanthropy amplify our efforts without burning out the team.

Rapid Feedback: Create systems that surface what's working and what isn't quickly enough to actually adjust course. This is a natural place for agentic philanthropy to log in. Evaluation and feedback at different milestones will help the team to iterate or course-correct along the way.

Collaboration isn't just about individual partnerships. It's about recognizing that every nonprofit, every donor, every volunteer is navigating the same **Global Charity Commons**. When one organization's cairn collapses, through scandal, mismanagement, or deception, it erodes trust across the entire landscape. When one builds with integrity, all benefit.

Ask Yourself

1. How might your organization reimagine its workspace, both digital and physical, to embody the values of connection and collaboration?
2. What projects would you want to propose be managed at the agentic island?
3. How might you personally use AI to amplify your work through the Agentic Island?

For Charities:

1. How might your charity use the Agentic Island to engage more staff and volunteers?
2. Is there one or two folks on your team that would enjoy being the first to use the Agentic Island?
3. How might you design an agentic island strategy that onboards new contributors (ie. charity autonomy) and fosters culture.

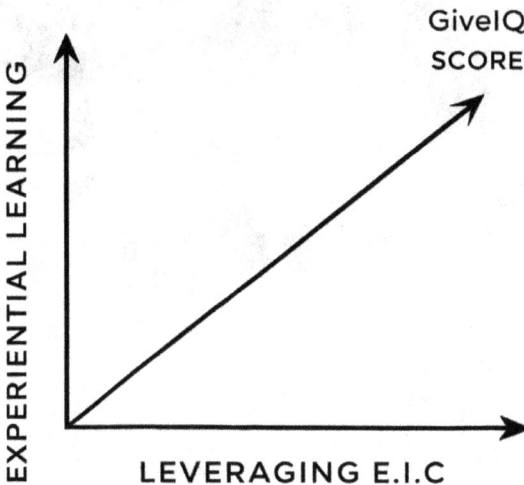

GivelQ
SCORE

EXPERIENTIAL LEARNING

LEVERAGING E.I.C

06

NEXT CHARITY: CHARITY 5.0

🔑 Key Takeaways

- ☑ **The EIC Framework:** Empower, Innovate, and Collaborate, three pillars rooted in ancient rhetorical principles (Ethos, Logos, Pathos) that provide stability amid technological disruption.

- ☑ **100% Mission Success:** The Carter Center's Guinea Worm eradication proves that complete mission achievement, not just incremental progress, is possible.

- ☑ **AI as the New Electricity:** Organizations that embrace AI wisely will unlock new opportunities; those that don't will fall behind.

Riprap & Resolve

Lessons from Experience: Giving Beyond Wealth

I remember how excited I was to learn about our first-time donors at AMC. These are the folks who had never donated to AMC and are taking a first step to contribute. It was a joy especially to see young families talk about how they worked together to decide on shared philanthropic goals. I was reminded of how excited I was when I was a first-time donor to the organizations I loved. My first donation was "micro" by most standards but that micro donation was monumental to my philanthropic journey and to how I've learned to include nonprofit organizations in my own planning.

I have been blessed to work with wealthy individuals over the years. Folks who had perhaps at one time given a micro donation and then continued to give at subsequent higher levels, or maybe they participated in a matching gift or a leadership gifts campaign. Perhaps their philanthropy included planning for the organization in their estate. These folks are not greedy people. Quite the opposite, they are humble people. Some of them started with nothing, all to see their hard work parlay into successful businesses or other valuable pursuits.

But here is the thing: most of the wealthiest people I have worked with gave when they were just starting out and didn't have much to give. They gave. They also volunteered. They got more engaged. The idea that they got rich first and then started giving is a myth. Giving when it's a stretch teaches you something that writing easy checks never can. It builds the muscle of generosity before the muscle of wealth. By the time they had significant resources, giving wasn't a new skill they needed to learn, it was already part of who they were. Many recognized their philanthropy – their time, talent, treasure and trust contributions – as a responsibility. Consequently, their charity and purpose path participation increased as their wealth increased.

Gus Jones' Purpose Path

Augustina "Gus" Jones perched on the colossal riprap boulders that embraced South Boston's Castle Island, the salty tang of the harbor air a familiar comfort. The 2030s pulsed with the hum of personalized transit. The era of sprawling parking lots had given way to agile, on-demand mobility, and the TrainPod was the quintessential last-mile solution to Castle Island's sunny harborside.

Emerging from Andrew Station's automated doors, Gus simply lifted her hand, the embedded chip in her palm a seamless identifier, and a sleek, ovoid TrainPod silently glided to her side.

Completing her circuit, Gus found her favorite spot on a massive, sun-warmed granite block within the riprap wall guarding the tranquil waters of Pleasure Bay. She inhaled deeply, savoring the revitalizing sea air, noticeably cleaner and crisper than even a few years prior.

Upon returning home, the subtle haptic feedback in her fingertips alerted her to two urgent messages from her friend Sam, currently stationed in Ethiopia. It wasn't an emergency but something far more exciting: a distinct, promising plan.

Sam's message articulated a powerful vision. He spoke of "Feudal AI," a burgeoning reality where powerful AI oligarchs controlled vast swathes of data and technology, creating new forms of digital inequality. This centralization had inadvertently spurred a counter-movement, driving many young, idealistic minds towards decentralized, community-owned alternatives.

His core idea was simple yet impactful: use blockchain technology to fund and manage decentralized water purification stations across underserved rural villages in Ethiopia. The project would be entirely crowdsourced, meticulously transparent, collectively maintained and governed directly by the communities it served.

A thrill of recognition shot through Gus; this was the very essence of what her philanthropic tech ideals aimed to foster, meaningful giving empowered by cutting-edge technology yet rooted firmly in the autonomy and needs of the communities themselves.

"Transparency and trust," Gus murmured, a fundamental tenet of GiveIQ echoing in her mind. The immutable ledger of the blockchain ensured that every financial transaction and every community decision would be permanently and openly recorded.

"But beyond the elegant technology, our true innovation lies in fostering genuine human connections. This is decentralized philanthropy in its truest form. Let's organize a virtual town hall with key Ethiopian community leaders next week. Their insights must shape every aspect of this project."

For Augustina "Gus" Jones, her path, her purpose, had never felt so clear.

The Emerging Tech Toolkit

Fortunately, the tools to support this new model of giving are not hypothetical; they're already emerging. The same technologies reshaping business and communication can be repurposed to empower everyday givers with unprecedented precision and reach. Consider that Lovable, an AI app-building platform, reached $100 million in annual recurring revenue in just eight months, representing what developers call the "vibe coding" revolution.[9] Replit's revenue exploded tenfold in nine months after releasing its AI agent. Organizations like Zinus are already using these tools to build internal systems in days rather than months, saving hundreds of thousands of dollars.[10] Now imagine leveraging these platforms not just for business but to rapidly prototype and deploy systems connecting specific community needs with willing givers in real time.

Picture using visual AI to create deeply affecting micro-documentaries for overlooked causes. Research shows video content is shared 1,200% more than text and images combined, and campaigns including video raise 34% more donations.[11] Tools like HeyGen and Synthesia (now used by more than 90% of the Fortune 100) enable nonprofits to produce professional videos in 140+ languages without cameras or editing expertise.[12] Consider how Google's NotebookLM, with its November 2025 "Deep Research" feature, could empower any donor to quickly digest complex impact reports. The tool can analyze up to 50 sources containing 500,000 words each and generate audio summaries in 50+ languages.[13] Google for Nonprofits now provides these capabilities free to charitable organizations, with groups like Infoxchange reporting they save a week's worth of time per project.[14]

9   Superframeworks, "10 Best AI Coding Tools 2025: Vibe Coding Tools Compared," Superframeworks Blog, November 8, 2025. https://superframeworks.com/blog/best-ai-coding-tools

10   Replit, "Replit vs Cursor: Which AI Coding Platform Fits Your Workflow?" Replit Discover, 2025. https://replit.com/discover/replit-vs-cursor

11   Resident.com, "Free AI Video Generator for Nonprofits: Tell Your Story Effectively," Resident.com Resource Guide, July 9, 2025. https://resident.com/resource-guide/2025/07/09/free-ai-video-generator-for-nonprofits-tell-your-story-effectively

12   Synthesia, "The 13 Best AI Video Generators (I've Actually Tested)," Synthesia, 2025. https://www.synthesia.io/post/best-ai-video-generators

13   Google, "NotebookLM Adds Deep Research and Support for More Source Types," The Keyword, November 15, 2025. https://blog.google/technology/google-labs/notebooklm-deep-research-file-types/

14   Google, "Google for Nonprofits Will Expand to 100+ New Countries and Launch 10+ New No-Cost AI Features," The Keyword, June 11, 2025. https://blog.google/outreach-initiatives/google-org/google-nonprofits-updates-june-2025/

These technologies represent a fundamental toolkit for democratizing and scaling purpose-driven action, fueling our "super charity agency." Radical philanthropy must evolve hand-in-hand with this burgeoning future of abundance. Proactively planning for a lifetime of purpose and generosity is not just noble; it's a crucial strategy. Adapting to these technological shifts makes this a seminal time for charities and givers alike, ensuring we don't miss the profound opportunity to contribute meaningfully at every stage.

Purpose Before Performance

This approach is grounded in a vital operating principle: purpose before performance. We aren't chasing growth merely for growth's sake. The objective is meaningful, measurable progress dedicated to solving humanity's most significant challenges. And in many of these challenges we can start thinking about how we can permanently solve them. This defines a clear strategic arc: it starts with focused improvement, deliberately creates feedback loops to build flywheel momentum, leverages AI and cultivates Charity Autonomy to scale impact intelligently, and diligently iterates, continually refining the system. Until the mission is fulfilled.

Our North Star must be the journey from better to an excellently completed mission. The goal isn't just to manage problems more effectively or mitigate their symptoms. The goal is to build systems so effective, autonomous, and replicable that the problem is fully solved. I know this is a tall order, but I truly believe that we are in position to solve some of the most intractable problems. I have had the honor to be in the room at so many discussions around framing charitable missions or developing strategic plans. I learned so much from folks who gave their ideas, expertise, musings and practical learnings. We are at a transformational time, when charities can move from strategies to deliver "better" to the ultimate work of declaring the mission "done." This is the profound purpose and potential of the next nonprofit.

Traditional nonprofit structures face significant limitations. High overhead costs, heavy dependency on donations, burdensome bureaucracy, and geographic constraints often result in inefficiencies and diminished impact. Despite best intentions, traditional charity models frequently struggle with a clear disconnect between effort and measurable outcomes. Notwithstanding how nonprofits manage to do so much despite budget limitations, this new age of agentic philanthropy open up doors that unlock the yoke of charity scarcity thinking.

To move beyond budget constraints and scarcity thinking, I think it's important for us to work with AI and technological tools that engage at an individual and personalized level. AI underlying empowering, innovating, and collaborating in the hands of individuals. Charity autonomy gives each individual a chance to step up when they want to and do more. Nonprofits that tap into this wave of agentic philanthropy will help individuals develop their personal giving identities. These agentic philanthropy driven organizations will be able to create purpose-aligned systems, that understand local conditions and human readiness.

The 100% Commitment

All the organizations I have been affiliated with, from staff positions to volunteer roles, have been constrained in some way in one or more of the following big general resource buckets of time, talent, treasure and trust. For instance, in two large digital transformations, including the implementation of customer relationship management software or CRM, those charities struggled under resource constraints. We did not have enough money to hire the higher level of implementation service (often a big upcharge in addition to the CRM software), needed for seamless integration. So that put the onus on staff to quickly get up to speed to roll out the new CRM. Subsequently, the chance for change narrowed around what one could afford and not how to achieve scale and realize the charity's full promise.

The adoption of a CRM is a good metaphor for how charities run today. The mainstream CRMs are very expensive and take years to get up and running. The bells and whistles they promise during the sales pitches are wonderfully detailed and seemingly can do it all. But in the end, it depends on the excellence of implementation and how the team uses the CRM daily to do their respective tasks. For the Charity 5.0 model, we can ask how we may get folks of their own volition to upgrade their skills and thereby have a greater mission impact. This is where the Charity 5.0 architecture needs to be designed to host a dynamic army of free agents, a personalized platform that gives each individual the tools to empower, innovate, and collaborate. But also operates in a personalized way to amplify one's talents and mind the gaps where we might be weakest.

Every charity, regardless of its size, faces persistent challenges that hinder its ability to maximize impact. As the first stepping stone, I believe nonprofits that embrace agentic philanthropy and the technological wave we are riding can achieve a 100% improvement in mission impact in five years. I know it's easy for me to write and difficult to implement, but I really believe that 100% improvement in five years is within reach of most nonprofits.

One amazing case study of "mission accomplished" is that of the African Guinea Worm eradication. The African Guinea worm causes a debilitating condition for those who have been plagued by it for thousands of years. Thanks to the Carter Center, the Gates Foundation, and a host of multinationals, the Guinea Worm disease has been eradicated.[15] A "100%" success versus a charitable cause making incremental progress.

Let's apply this 100% success to every mission out there. First, identify what is holding the organization back from achieving greater success. As Andrew Ng said, AI is the new electricity.[16] For those who use AI wisely, including around the GiveIQ principles of 'empower, innovate, and collaborate'—it will unlock new opportunities.

Qualia and the Human Foundation of GiveIQ

GiveIQ begins with a simple conviction: intelligence without experience is incomplete.

Philosophers call this lived experience qualia, the felt reality of being human. It's the difference between knowing what a sunset is and standing still long enough to feel it. Qualia is where meaning forms. It's also where purpose begins.

---

15   Carter Center, "Guinea Worm Eradication Program," Atlanta: The Carter Center, 2024. https://www.cartercenter.org/health/guinea_worm/index.html

16   Andrew Ng, "AI Is the New Electricity," Keynote address at Stanford Graduate School of Business, 2017.

I've spent thirty years in the nonprofit sector, and I can tell you that the most important decisions I've ever made weren't driven by data alone. They came from standing in a half-built Habitat home, or sitting across local residents who told me our carefully designed parks program wouldn't work. From watching my mother's face when we moved her into memory care. Data informed those moments. But qualia, the lived, felt experience of being there, shaped what I did next.

Aristotle understood this intuitively. His triad of Ethos, Logos, and Pathos describes not abstract concepts, but human capacities shaped by experience. Ethos is character earned over time. Logos is judgment applied to real-world conditions. Pathos is empathy born from shared understanding. None of these exist without lived experience.

GiveIQ carries this foundation forward through a modern operational lens: Empower, Innovate, Collaborate.

Empower
Translates Ethos into agency rooted in responsibility

Innovate
Translates Logos into experimentation guided by judgment, not optimization alone

Collaborate
Translates Pathos into trust built through shared experience and dignity

AI can assist logic. It can simulate language. It can optimize systems. But it cannot generate qualia. It does not feel consequence, responsibility, or care.

For this reason, GiveIQ does not attempt to replace human judgment. It is designed to protect it.

The purpose of this framework is not to make philanthropy faster or more efficient at any cost. It is to ensure that, as tools grow more powerful, human experience remains at the center of how purpose is discovered, decisions are made, and impact is measured.

Know Thyself: The Hardest Part of the Purpose Path

Qualia grounds us in experience. But experience can also deceive us.

We all approach things with our own self-confidence, and sometimes blind spots, thinking we can make a stronger contribution than what's actually possible. I've had to let some people go over the years, both staff and volunteers. Some have been incredulous. "I'm a volunteer making a contribution," they thought, "and now they're saying they don't need me? How can they not need a volunteer?" Or a staff member who thought they were irreplaceable, including a few who assumed, "Hey, this is a charity, they don't fire people."

But along the way, there are places that aren't good fits. Including for myself. Situations where the culture, the people, the mission, one's own skills, one's own vision for what needs to be a priority, none of it equates with what's actually happening on the ground.

I've seen it from every angle. The donor who wanted to give a large gift to fund a specific project, and we had to say that wasn't a priority for the organization. The employee who felt they were the consummate expert in a specific vertical, how could we question them? I've heard twenty-eight-year-olds tell me they were the consummate expert on nonprofit management, despite having only worked in one vertical of the business. These are people who knew one particular area of an organization pretty well and had held one job. How could they be the modern equivalent of a hero who could do everything, know everything, at an expert level?

I still think, even after thirty-five years, that I have so many blind spots when it comes to running an organization, understanding that organization's culture, and knowing in what areas my contribution is actually relevant and strong.

This is where honest self-assessment matters. Not the self-assessment that flatters, but the kind that reveals where you're genuinely valuable and where you're not. Here's where the "T" for trust emerges. The GiveIQ Score is designed to help surface these patterns, not to shame, but to guide. Where are you strong? Where do you overestimate your contribution? Where might someone else be a better fit?

And this is where AI can amplify our strengths and help compensate for our weaknesses. It can also bring in that outsider resource, the person or skill that won't happen through serendipity but now, through agentic philanthropy, connects and matches us at a global scale. The twenty-eight-year-old who thinks they know everything about nonprofit management might actually have deep expertise in one specific area. AI can help match that expertise to where it's genuinely needed, rather than letting overconfidence place them where they'll struggle.

Know thyself isn't just ancient wisdom. It's operational necessity. And in the Global Charity Commons, honest self-knowledge becomes a gift to everyone, because when you know where you fit, you stop blocking the path for others who fit better.

The EIC Framework: Empower, Innovate, Collaborate

The GiveIQ model is built on three forms of human agency: the agency to act, the agency to create, and the agency to connect. We call these Empower, Innovate, and Collaborate.

In my own nonprofit career, through years working with charities of all sizes, these are the areas I've identified as the ones that unlock both individual and organizational promise. They're where potential becomes real.

And these three pillars aren't arbitrary. They're rooted in the ancient rhetorical principles that have guided human persuasion and action for millennia:

Empower builds on Ethos, the credibility that comes from character and capacity

Innovate applies Logos, the power of reason and evidence to solve problems systematically

Collaborate taps into Pathos, the emotional and relational bonds that move people to act together

This connection to timeless principles matters because EIC provides stability in an era of constant technological disruption. Nonprofits adopting ChatGPT today may face advancements like Google Gemini or xAI Grok tomorrow. Coding tools like Lovable. dev and Replit add new capabilities daily. The cost of AI intelligence is plummeting. Soon we'll have access to nearly all of these tools for little cost.

The tools will keep changing. EIC won't.

EMPOWER: The Agency to Act

Empower is rooted in Ethos. Not just reputation, but the credibility that comes from who you are and who you're becoming. Your character. Your capacity. Your readiness to act effectively in the world.

At its heart, Empower means using AI to expand human potential, not replace it. Imagine nonprofits deploying AI-assisted training programs that offer personalized coaching. AI-powered volunteer engagement platforms that match people to roles where their skills and passions deliver the greatest impact.

The global health nonprofit PATH offers a compelling example. By using AI-driven digital health assistants, they provide frontline healthcare workers in resource-limited settings with real-time medical guidance. A community health worker in a remote village, facing a complicated case, can access expertise that would otherwise require years of training or a specialist hundreds of miles away.

The AI doesn't replace the healthcare worker's judgment. It amplifies their capacity to act. That's Empower.

INNOVATE: The Agency to Create

Innovate draws on Logos. The power of reason and evidence to solve problems systematically. But reason without creativity is just analysis. Innovate is about bringing your full creative capacity to challenges that old thinking hasn't solved.

This means applying emerging technologies in strategic, mission-aligned ways. AI-powered donor segmentation and predictive analytics make hyper-personalized outreach possible, connecting donor passions with specific needs in ways no human team could scale.

DonorsChoose exemplifies this principle. They use AI-driven recommendations to match individual donors with classroom projects that align with their interests and giving history. A donor passionate about STEM education in rural areas sees projects that match that passion. The result: significantly higher funding success rates and deeper donor engagement.

The AI doesn't replace human creativity. It opens possibilities humans hadn't seen. That's Innovate.

🔳 COLLABORATE: The Agency to Connect

Collaborate taps into Pathos. The emotional and relational bonds that move people to act together. But connection without coordination is just community. Collaborate is about building with others toward outcomes none of you could achieve alone.

AI can serve as a powerful catalyst for this kind of partnership, enabling organizations to identify shared opportunities, optimize resource allocation across networks, and co-develop solutions that no single organization could create.

Ushahidi, the nonprofit crisis-mapping platform born in Kenya, demonstrates what becomes possible when the agency to connect is amplified at scale. During the 2010 Haiti earthquake, Ushahidi's platform enabled nearly 40,000 independent reports to flow in from affected communities via text message and social media. Volunteers around the world collaborated to verify, translate, and map these reports in real time. The result was what FEMA's director called "the most comprehensive and up-to-date map available to the humanitarian community." A seven-year-old girl and two women were pulled from rubble after responders received their text message through the platform.

Today, Ushahidi is incorporating AI to categorize, prioritize, and interpret the massive amounts of incoming data during crises, making it possible for communities to coordinate response faster than ever before. The platform has now been used in over 160 countries for everything from election monitoring to COVID-19 response.

The AI doesn't replace human connection. It amplifies our capacity to find each other, coordinate action, and build collective intelligence. That's Collaborate.

Three Forms of Agency, One Framework

The agency to act. The agency to create. The agency to connect.

These aren't new. They're as old as Aristotle. What's new is the amplification.

Agentic philanthropy puts AI alongside your agency, not in place of it. PATH's healthcare workers are still making the decisions. DonorsChoose donors are still choosing. Ushahidi's volunteers and affected communities are still the ones raising their voices and coordinating response.

The tools will keep evolving. New platforms will emerge. Costs will continue to fall. But EIC remains constant: Empower, Innovate, Collaborate. The agency to act, create, and connect.

That's the foundation everything else builds on.

The Power of Listening: Co-Creating Programs

In 1996, I embarked on a new journey in the nonprofit sector, transitioning from a rewarding six-year career in Scouting. During this time, I found myself working with Morgan Memorial Goodwill Industries.

As Director of Corporate and Foundation Relations, I had the privilege of collaborating with Mary Reed, Goodwill's Vice President, overseeing educational and training programs. Our shared mission was to develop an educational program tailored specifically for urban teenage girls.

Our team spent weeks brainstorming ideas and logistics. Eventually, we landed on what seemed like a logical solution: an after-school weekday program. But before we finalized anything, we made the critical decision to consult the very people we intended to serve, the girls themselves.

In a pivotal listening session, we gathered a group of teenage girls at the Goodwill Summer Camp in Athol, Massachusetts, and presented our plan. What followed was both enlightening and humbling. The young women explained in detail why our afternoon program was unworkable. Some were latchkey kids responsible for caring for younger siblings, others had part-time jobs, and many were committed to sports and extracurricular activities.

Just as Mary and I exchanged uneasy glances, worried that we would need to start from scratch, one of the teens offered a solution: "What about Saturday mornings?" Heads nodded in agreement.

Their input changed everything. What emerged was the Saturday Academy, funded generously by Mellon Bank and championed by their philanthropic director, Joan Jaxtimer. This program, shaped by the voices of its participants, flourished and continues to do so to this day.

Had we not engaged in that moment of listening, we might have built something that was well-funded but ultimately ineffective.

Listening Today: Amplified by AI

Today, AI holds incredible potential to supercharge our ability to listen at scale. AI-powered tools can sift through massive volumes of feedback from diverse sources. Natural Language Processing allows us to analyze this input for sentiment, recurring themes, pain points, and unspoken needs. And personally, AI offers us a chance to listen to ourselves – to "know thyself" - and thereby help to distinguish from the trivial to what truly matters.

By combining the age-old wisdom of listening with the power of AI, we can build smarter, more responsive social sector solutions; designed with, not just for, the communities we serve. This gives us greater assurances on funding and programs operating at the edge – perhaps in a distance village or place we will never physically visit.

AyitiDAO: Decentralized Humanitarian Relief. A fictional case study.

AyitiDAO puts control in Haitian hands. Women farmers in Artibonite vote on which projects receive funding. Smart contracts release funds only when real-world milestones are validated. The EIC principles in action: Empower local communities, Innovate with blockchain accountability, Collaborate with on-the-ground NGOs.

How It Works: Organizations submit proposals → Community votes on funding → Smart contracts release funds as milestones are met → Contributors earn reputation for delivery.

Why This Model Is Powerful:

- ❑ It replaces centralized gatekeeping with community-led decision-making
- ❑ It uses smart contracts to enforce real-world performance, reducing corruption
- ❑ It offers transparent accounting, accessible to all
- ❑ It enables Charity Autonomy, where participants are not passive beneficiaries but active designers of their own future

The 4 T's: Operational Fuel for Charity 5.0

❑ The 4 T's—Time, Talent, Treasure, and Trust—aren't just operational categories. They're the architecture that allows nonprofits to move at velocity while staying anchored to mission. And they're how individuals exercise their agency to act, create, and connect.

❑ Here's how each T transforms in the Charity 5.0 era:

	TODAY	CHARITY 5.0
TIME	Scheduled volunteer shifts, event-based engagement, annual commitments	Flexible micro-contributions, AI-optimized scheduling, asynchronous participation, 90-day Agentic Islands
TALENT	Hiring staff, recruiting volunteers, occasional pro bono projects	Dynamic talent ecosystems: humans and AI agents working together, skills-matched assignments, expertise on demand
TREASURE	Campaigns, donor tiers, seasonal pushes, capital projects, annual fund appeals	Generative wealth: AI-powered personalized giving, blockchain transparency, pooled giving circles, crypto-donations, smart contracts
TRUST	Board oversight, annual reports, third-party ratings, centralized compliance	Distributed integrity: peer validation, blockchain tracking, real-time impact documentation, network activation, recruiting and connecting others

❏ **Time** shifts from scheduled blocks to flexible engagement. Instead of asking "Can you volunteer every Tuesday?", Charity 5.0 asks "What 15 minutes do you have right now?" AI-optimized scheduling matches contributor availability to organizational needs in real time. Agentic Islands provide bounded 90-day commitments that respect busy lives while enabling meaningful contribution.

❏ **Talent** evolves from hiring and recruitment to dynamic ecosystems. The question isn't "Who can we get on staff?" but "What combination of human expertise and AI capability solves this problem best?" Skills-based matching connects specialized knowledge to specific needs, whether that's a retired logistics manager redesigning a food bank's distribution system or an AI agent handling donor communications at scale.

❏ **Treasure** expands beyond traditional fundraising. Generative wealth means AI-powered personalization that connects donor passions to specific opportunities. It means blockchain transparency that lets contributors see exactly where their dollars go. It means giving circles that pool resources for collective impact. The shift is from extracting donations to co-creating philanthropic value.

❏ **Trust** transforms from centralized verification to distributed integrity. Instead of waiting for annual reports and third-party ratings, trust builds through real-time impact documentation, peer validation, and transparent tracking. When Keisha recruits five friends to volunteer, their confirmation of her role strengthens the system's credibility. When a nonprofit's outcomes are tracked on a shared ledger, trust becomes verifiable rather than assumed.

❏ The 4 T's remain the same resources they've always been. What changes is how we mobilize them. In Charity 5.0, each T becomes more fluid, more connected, and more powerful through the amplification of agentic philanthropy.

EIC: The Engine of Value Creation

Culture is the mindset. EIC is the method.

EIC VALUE FLYWHEEL

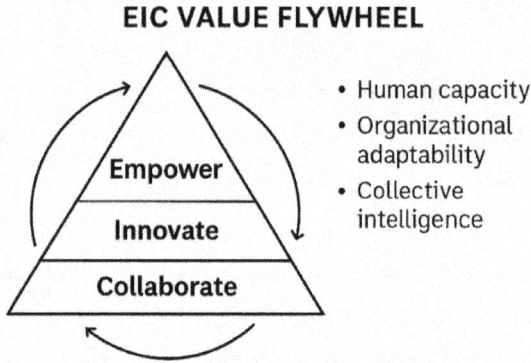

- Human capacity
- Organizational adaptability
- Collective intelligence

Empower

Innovate

Collaborate

Empower develops human potential.

Innovate transforms capacity into new possibilities.

Collaborate amplifies and scales outcomes.

The Next Charity Model: Charity-as-a-Service

With personalization comes the need for on-demand infrastructure. We move beyond the concept of the charity as a fixed institution towards a frictionless, on-demand philanthropy-as-a-service ecosystem. Imagine an AI-driven request-response system akin to Uber, connecting those who need help with those who can provide it in real-time, with transparency and accountability built in. Charities evolve into dynamic impact marketplaces where individuals, nonprofits, and businesses can seamlessly plug in to contribute and collaborate. There are many charities now that are sharing back-office services. We need to scale the concept of back-office service sharing and subsequently enable charities to focus on what they are best at.

In the age of AI, philanthropy doesn't have to wait for a formal appeal, it can be embedded in the everyday. Technology now enables seamless, spontaneous giving through micro-donations triggered in the flow of daily life. These include automated round-ups at the point of sale, AI-driven contributions aligned with a donor's values, and even social tipping for verified acts of kindness. For instance, while my family and I visited the Monterey Bay Aquarium we heard their trained staff ask if we'd like to "round up" on our purchases at the gift shop or café, channeling spare change into marine conservation efforts. This strategy, now replicated by other nonprofits and retail partners, demonstrates how small gestures, when aggregated by sophisticated AI and financial tools, can generate meaningful impact. It illustrates the distributed power of generosity at scale, turning fleeting moments into sustained contributions for good.

A blockchain-based reputation system can meticulously track and reward volunteering efforts and demonstrated impact, creating a verifiable "Impact Score" for every individual. This portable ID recognizes contributions across organizations, fostering a culture of sustained engagement and providing a powerful incentive for pro-social behavior. This empowers folks who lack, for a better word, a credential, an impact credential that one can take and use to open doors to do good work elsewhere, including around the globe.

This first-principles lens reveals a new architecture for charity, one that is agile, transparent, and relentlessly focused on outcomes.

Decentralized Impact Networks (DINs)

Replacing hierarchical organizations with dynamic, interconnected Impact DAOs. Funders, volunteers, and beneficiaries engage directly. Smart contracts automatically disburse funds upon the verifiable achievement of pre-defined impact milestones.

Precision Philanthropy

AI evolves into a personalized philanthropy concierge, guiding individuals and organizations to their optimal "Purpose Path." Dynamic AI assistants autonomously allocate donations, responding in real-time to evolving needs.

Quantifying and Incentivizing Good

Charity Tokens or Impact Credits provide a tangible mechanism to quantify and track contributions and their resulting impact. Blockchain technology ensures that every donation and volunteer hour is verifiable, traceable, and auditable.

The Digital Transformation Thesis

A New Design Principle for Philanthropy

What if philanthropy worked more like the human brain?

Not as a metaphor, but as a design principle.

The human brain does not rely on a single master neuron. It functions because millions of individual neurons act independently, each contributing when they reach their own threshold, forming patterns of intelligence that are adaptive, resilient, and remarkably efficient. No central authority dictates every decision. Intelligence emerges through participation.

This book introduces that same principle to philanthropy.

For the first time, philanthropy is reframed as a neuromorphic system: a distributed network of individuals, communities, organizations, and intelligent agents, each capable of autonomous action, learning, and contribution. When enough participants engage, collective intelligence forms. When participation is widespread, impact compounds. Just as neuromorphic chips mimic how neurons fire and connect in the brain, we're creating a charitable system where individual givers spark connections that build into something more intelligent than any single donation. This is not an argument against institutions. It is an argument against concentration.

For too long, philanthropic decision-making has been constrained by gatekeepers, intermediaries, and centralized models that unintentionally limit agency. While these systems were designed for efficiency and accountability, they often reduce participation, slow adaptation, and narrow the definition of who gets to contribute and how. They added friction and oftentimes created closed spaces of inclusivity. A neuromorphic approach offers a different path forward.

At the heart of that path is Charity Autonomy. As you have read, Charity Autonomy reflects a fundamental shift in how purpose and philanthropy are activated. Individuals are no longer limited to passive roles as donors, volunteers, or beneficiaries waiting to be invited into a larger effort. Instead, they are empowered to act, learn, and contribute directly. Charity Autonomy makes philanthropy something that can be engaged on demand, allowing purpose and service to become a more constant presence in our lives rather than an occasional or episodic act. It recognizes that human agency, creativity, and willingness to help already exist. What has been missing is a system designed to let that agency surface.

GiveIQ™

AI Agent Glyph™
indicating presence
of verified AI systems

Agentic philanthropy is what makes Charity Autonomy possible at scale. It is the AI-enabled layer that works alongside human judgment to support, coordinate, and accelerate charitable action. Agentic philanthropy does not replace human values or decision-making. It amplifies them. AI agents help participants recognize patterns, surface opportunities, reduce friction, and manage complexity, allowing people to focus on what they do best: imagining solutions, exercising judgment, and taking action. Together, Charity Autonomy and agentic philanthropy form the operational foundation of a neuromorphic philanthropic system, where intelligence emerges through participation rather than control.

This foundation comes to life through Agentic Islands. Agentic Islands are focused projects and initiatives where people come together around a shared purpose, supported by AI agents designed for the specific tasks at hand. These islands create space for creativity and experimentation, allowing ideas to be explored, refined, and stress-tested through digital twins before being executed in the real world. Rather than waiting for hierarchical approval or institutional sign-off, participants can move from insight to action more quickly, learning in the open and improving as they go. Agentic Islands operate at human scale, but they are designed to connect.

Over time, those connections give rise to the Global Charity Commons: a shared philanthropic ecosystem where knowledge, trust, learning, and impact persist beyond any single organization, campaign, or moment. In the Commons, progress compounds rather than resets. What is learned in one place can inform action elsewhere. What works once does not need to be reinvented repeatedly. As this ecosystem matures, it creates the conditions to meaningfully move the needle on some of the world's most pressing challenges, not through centralized authority, but through coordinated, intelligent action at scale.

One of the core premises of this book is that this shift dramatically raises what is possible. With widespread Charity Autonomy, supported by agentic philanthropy and connected through the Global Charity Commons, charities should be able to achieve a 100 percent improvement in impact over the next five years. And in time, for certain causes, something even more ambitious becomes imaginable: 100 percent mission success. The near-eradication of Guinea worm disease offers a glimpse of what this future can look like. GiveIQ argues that by designing philanthropy around human agency and collective intelligence, outcomes like this no longer need to be rare exceptions. They can become achievable goals.

This shift marks the true digital transformation of philanthropy. Not replacing human values, but amplifying them.

The equation remains constant: Human agency plus agentic philanthropy equals greater societal impact. Charity 5.0 is simply the infrastructure that makes that equation scale.

Ask Yourself

For Nonprofit Leaders:

1. How might your organization reimagine its current workspace to embody connection and collaboration?
2. In what ways can your team balance the autonomy of AI agents with human oversight to ensure alignment with your core values?
3. What funding and other resources are needed to ensure success?

For Anyone Building Something New:

Forget your current systems, culture, and constraints for a moment. If you were launching your nonprofit today as a startup, what would you build? What technology would you use? How would you structure decision-making?

Now ask: which of those innovations could you actually implement in the next 90 days?

DIGITAL TWIN

PREDICTIVE AI, DIGITAL TWINS & CYBERSECURITY

🔑 Key Takeaways

- ☑ **Digital Twins for Impact:** Virtual replicas updated with real-time data enable nonprofits to simulate scenarios, optimize resources, and make better decisions before committing resources.

- ☑ **Unstructured Data Is Gold:** AI can transform decades of overlooked archives, volunteer notes, and citizen-generated data into actionable insights.

- ☑ **Cybersecurity Is Mission-Critical:** As charities grow in importance and wealth, they become cyber targets; robust security underpins effective governance and donor trust.

The Visionary Journey of Gus Jones

Gus Jones had always been a maverick, a trailblazer determined to harness data for transformative urban planning. In 2050, as Lagos endeavored to reinvent itself from a crowded metropolis into an Outdoor City, Gus answered the call. The Lagos government had commissioned her team to help implement an ambitious **Outdoor City** project that transformed neglected urban spaces into vibrant, accessible green areas.

Lagos, home to nearly 20 million people, had long been hampered by development policies that prioritized buildings over nature. In an inspiring move, the city set out to correct these errors by designing a healthier, happier urban environment where green spaces would be woven into the fabric of everyday life.

Gus and her team embarked on an ambitious project to create an Outdoor City digital twin of Lagos. This dynamic, interactive, virtual replica could simulate how new green spaces would interact with existing urban structures. Stepping into a fully immersive simulation chamber, Gus found herself before a 3D holographic model of Lagos. Here, predictive AI tools enabled a "what if" laboratory: every query began with "what if?" as human experts and AI agents explored countless scenarios.

In the digital twin simulation, every "what if" scenario was meticulously explored. Planners could see how green spaces would extend to underserved neighborhoods, how these spaces might mitigate urban heat islands, and even how they would behave under the strain of natural disasters.

Gus's pioneering work reimagined urban spaces and set the stage for a future where data-driven insights and immersive simulations would become the norm. The digital twin gave Gus and others an eye of what Lagos could become, including how to engage generations in the outdoors and conservation stewardship.

Rediscovering Unstructured Data

In every organization, vast amounts of data accumulate over time. Historically, much of this data has been unstructured, stored in file cabinets, or relegated to forgotten archives. Yet, the advent of AI and advanced Visual Language Models is breathing new life into this overlooked resource.

When I began my work at the Appalachian Mountain Club (AMC) in 2012, we were at the start of a digital transition. For decades, volunteers would head out into the woods with tiny golf-sized pencils and paper, eventually "waterproofed" inside a plastic sandwich bag.

I recall many conversations with conservation experts who debated the validity of citizen-generated data. Many were skeptical, insisting that the inherent inconsistencies in data collected by laypeople rendered it too unreliable for rigorous analysis.

Fast forward to today: the evolution of technology has transformed our approach. Where once the "three P's" (people, paper, and pencils) were the norm, we now use smartphones, cutting-edge applications like iNaturalist,[17] and sophisticated AI tools to validate and analyze every piece of information. Data that was once dismissed as error-prone is now refined into statistically significant insights.

---

17   iNaturalist, "Citizen Science for Biodiversity," California Academy of Sciences and National Geographic Society, 2024. https://www.inaturalist.org

Digital Twins in Action

Digital Twins: From Science Fiction to City Planning

Digital twin technology is no longer experimental. According to ABI Research, cities are expected to save more than $282 billion annually by 2030 through digital twin implementations. Singapore's Virtual Singapore project has created a comprehensive 3D digital twin of the entire city, enabling real-time simulations of everything from traffic flow to disaster response. Cities like Helsinki, Boulder, and Sacramento are already using digital twins for urban planning, infrastructure monitoring, and community engagement.

In the Spring of 2023 I was invited to speak at Singapore's Urban Redevelopment Authority. The bringing together of data science, design principles and a cultural prioritization were world class. Upon entering the Urban Authority's space I spent as much time as I could walking around their giant 3D city model. It is complete with recent construction and parks projects and those underway. It provides a stunning three dimensional visual of Singapore and what has been accomplished in less than sixty years.

For nonprofits, a 3D model like Singapore's creates a unique touch point to look back, teach others, and a model for planning the future. The emerging digital technology offers unprecedented capability. Imagine a food bank using digital twin simulations to optimize distribution routes before deploying trucks. Or a housing nonprofit modeling the impact of different development scenarios on community wellbeing. The technology that once required major corporate investment is becoming accessible to mission-driven organizations of every size.

A digital twin is a virtual replica of a physical object or system that is continuously updated with real-time data. More than a mere copy, a digital twin is an evolving simulation that mirrors the state, behavior, and nuances of its physical counterpart.

Digital twins offer:

- ❑ **Risk mitigation:** Simulating scenarios to foresee potential problems before resources are spent
- ❑ **Resource optimization:** Determining the most effective ways to allocate time, money, and effort
- ❑ **Enhanced decision-making:** Offering a dynamic, 360-degree view of complex systems

Cloud computing platforms such as Amazon Web Services have begun democratizing access to digital twin technology.[18] Advances in computational power mean that even organizations with modest budgets can now harness these powerful tools. Moreover, as we introduce agentic tools we will be able to solve problems that now seem intractable.

Cybersecurity for Nonprofits

Years ago, I learned firsthand just how ever-present cyber threats can be. I received an email from a junior staff member on our purchasing team. The staffer was following up to confirm that he had gone ahead and purchased the Apple gift cards, as I had supposedly "instructed," and asked if I wanted the physical cards mailed to me. I immediately replied, telling him I had never requested any Apple cards. Unfortunately, by then, it was too late; he had been duped. The scammers had convinced him to email the digital card codes, and just like that, $1,800 was gone without a trace.

My experience was far from unique. According to the Federal Trade Commission, gift card scams alone resulted in $217 million in losses during 2023, with nearly 20,000 complaints representing $100 million in additional losses in just the first half of 2024. [^32] Nonprofits have become prime targets: 60% have reported experiencing a cyberattack in the last two years, and the sector saw a 30% year-over-year increase in weekly cyberattacks in 2024.[^33] Email threats against nonprofits rose 35%, with credential phishing surging over 50%.[^34] The consequences can be devastating.

---

18   Amazon Web Services, "AWS IoT TwinMaker: Digital Twin Technology," Seattle: Amazon Web Services, 2024. https://aws.amazon.com/iot-twinmaker

Save the Children International lost 6.8 terabytes of data to ransomware in 2023. The International Committee of the Red Cross saw personal information of 515,000 vulnerable people compromised in 2022.[^35]

That experience was a wake-up call. In response, we swiftly implemented cybersecurity training for our employees. As one might expect, the first phishing email we sent as a test, our employees failed. Many were duped into believing it was real. With each iteration they improved in their awareness. This is critical work: nine out of ten nonprofit organizations do not train staff regularly on cybersecurity, and 80% don't have a policy in place to address cyberattacks.[^36]

As a sector, charities must work to build a vibrant cybersecurity ecosystem. Establishing cyber and agentic guardrails and new perimeters, while keeping human agency in control. They should:

- ❏ Engage partners in industry, academia, and government
- ❏ Conduct risk assessments that protect critical assets
- ❏ Ensure compliance with cyber regulations
- ❏ Implement cyber awareness training

❝ ────────

"Policies toward escalation and de-escalation for the physical world need to be extended to the digital world if cyberspace isn't going to be the 'Wild West.'"

NATE FICK,
FORMER U.S. AMBASSADOR AT LARGE FOR
CYBERSPACE AND DIGITAL POLICY[19]

──────── ❞

───────
19   Nate Fick, Remarks as U.S. Ambassador at Large for Cyberspace and Digital Policy, Washington, D.C., 2024.

AI Risks in Nonprofits

Artificial intelligence is a force multiplier for philanthropy. However, the inherent risks of AI-generated misinformation pose serious challenges. As AI-generated content proliferates, ensuring accuracy and maintaining trust in nonprofit communications becomes paramount.

Research shows that AI, through multi-agent frameworks, can act as a frontline filter, automatically catching up to 90% of errors. This allows human reviewers to focus on the remaining issues, making the overall review process more manageable while safeguarding brand integrity.

Organizations must be proactive. AI systems should be built with preventative mechanisms that integrate real-time hallucination detection before content goes live or critical decisions are made.

Innovation that extracts value without strengthening cairns degrades the **Global Charity Commons**. True innovation adds wisdom/knowledge stones that others can build upon, open-source tools, shared learning, transparent failures that save others from the same mistakes.

The Human in the Loop

AI processes data, identifies patterns, simulates scenarios. Humans decide what matters. You're stewards of public trust. The efficiency gains from AI can never come at cost of accountability. Stay in the loop.

Ask Yourself

For You Personally:

1. **Understand Your Data Footprint.** You're already generating data constantly: spending patterns, health metrics, learning habits, giving history. If you could see all that data mapped out, what patterns would surprise you? What intentional changes would you make? Are you ready to see yourself that clearly?

2. **Simulate Before You Commit.** Before you make your next major decision, career change, major donation, volunteer commitment, would you want to see three different scenarios play out in a digital twin? What would you want to test? What outcomes would you measure?

3. **Protect What Matters.** Who has access to your giving data right now? Your health information? Your location history? Do you know? If a nonprofit asked for permission to use your anonymized data to improve their programs, what safeguards would you need to see before you'd say yes?

For Nonprofit Leaders:

1. **Know What You're Sitting On.** Pull up your donor database, your volunteer records, your program outcomes data. Now ask: What patterns are invisible to you right now that a digital twin could reveal? Which volunteers are at risk of disengaging? Which donors are ready to increase giving? Which programs are succeeding for reasons you don't understand yet?

2. **Test Before You Launch.** Your board wants to launch a new program, open a new location, or pivot strategy. Before you commit resources, could you simulate it? Run it through a digital twin for 90 virtual days. What would you want to measure? What failure modes would you want to test? What would you learn that could save you from an expensive mistake?

3. **Secure Your Digital Infrastructure.** If your systems were compromised tomorrow, what donor data would be exposed? What program operations would halt? What's your actual cybersecurity posture versus what you tell your board it is? When was the last time you ran a penetration test?

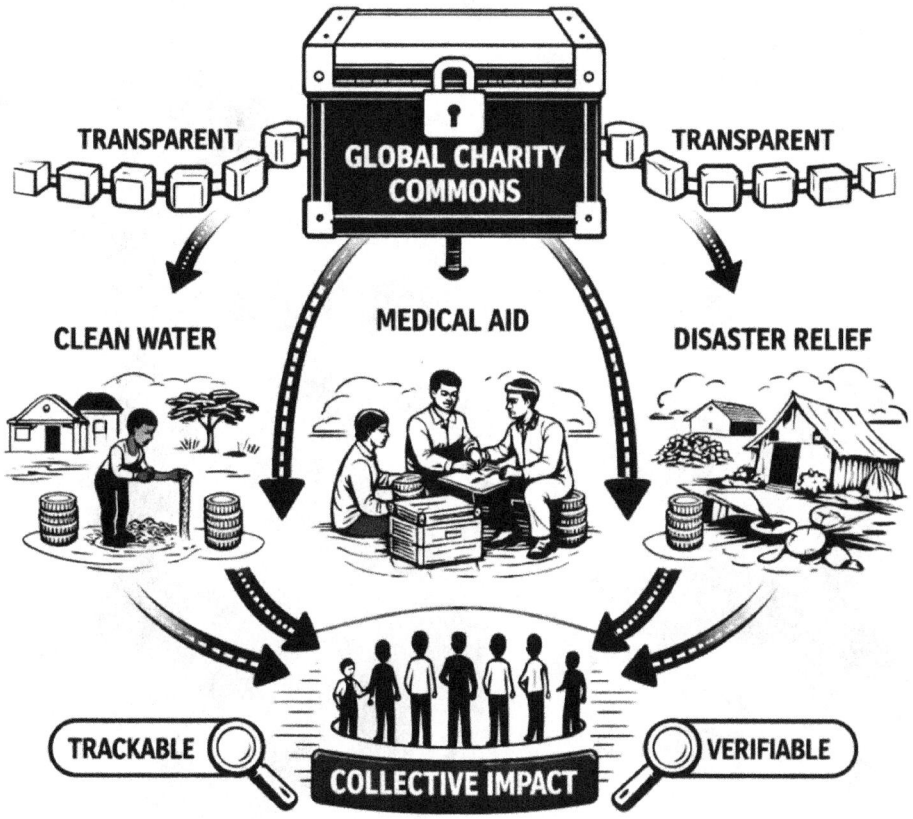

08

WEB 3.0 & DECENTRALIZATION

🔑 **Key Takeaways**

☑ **Web 3.0 Enables Transparent Philanthropy:** Smart contracts, DAOs, and blockchain create trustless systems where every donation is traceable and fund flows are transparent.

☑ **The Global Charity Commons:** A shared civic cloud where nonprofits, donors, volunteers, and AI agents can exchange insight and coordinate action in real time.

☑ **From Parallel Trails to Collective Maps:** When individual Purpose Paths become visible, disconnected efforts can converge into coordinated collective impact.

Blockchain's transformative potential in philanthropy goes beyond simple transparency. Consider the possibilities: Every donation becomes a verifiable digital asset. Impact metrics are recorded immutably. Donor intent can be encoded into smart contracts that execute automatically when conditions are met. A donor could specify that their gift only releases when a nonprofit achieves certain outcomes, and the system would enforce that automatically, without intermediaries.

This isn't about replacing trust with technology. It's about creating infrastructure that makes trust easier to build and harder to abuse. When a nonprofit's financial flows are visible on-chain, the conversation shifts from "prove you're legitimate" to "let's talk about impact." When volunteer hours are cryptographically verified, reputation becomes portable across organizations and borders.

The Global Charity Commons emerges from these building blocks. It's not a single platform but an interconnected ecosystem, think of it as the philanthropic equivalent of the internet itself. Just as the internet enables anyone to publish content and connect with anyone else, the Global Charity Commons enables anyone to contribute to causes and connect with impact opportunities worldwide. The barriers aren't technological anymore. They're cultural and institutional. And those are exactly the barriers that GiveIQ is designed to break down.

Unlocking Impact with Web 3.0

We stand at a unique moment where individual purpose, collective coordination, and planetary problem-solving can finally converge. Web 3.0 is more than a tech shift; it's a philosophical change rooted in decentralization, transparency, and user ownership. Chris Dixon describes Web 3.0 as the "ownership web,"[20] with technologies such as smart contracts, DAOs, tokens, and decentralized apps enabling new ownership and governance.

Blockchain's Transformative Potential in Philanthropy:

❑ **Smart contracts** enable automated, transparent distribution of funds

❑ **DAOs** empower communities to manage philanthropic efforts collaboratively

❑ **Tokenizing contributions** allows donations to become tradable, traceable digital assets

In Web 3.0 contexts, cairns become virtual but no less critical. Every review you leave, every impact story you share, every transparent assessment of a nonprofit's effectiveness—these are digital stones that guide others navigating the Global Charity Commons. The same principles apply: build on solid ground, be honest about what works and what doesn't, strengthen the path for those who follow.

The Cairn Chain: Permanence in a Disposable World

Chris Dixon calls it the "ownership web." I call it something simpler: building cairns.

If you've ever hiked above treeline, you know what a cairn is—those carefully stacked stone markers that guide you when there's no trail to follow. Some passing hikers may take the time to add a stone. Nobody owns the cairn. Everybody benefits from it. And it endures long after any individual hiker is gone.

---

20   Chris Dixon, *Read Write Own: Building the Next Era of the Internet* (New York: Random House, 2024).

That's what blockchain technology makes possible for philanthropy. Not the speculative cryptocurrency nonsense that dominates headlines, but something more profound: a permanent, immutable record of charitable action. Each donation, each volunteer hour, each verified outcome becomes a stone in a shared cairn—visible to all, owned by none, guiding those who come after.

Stanford's Graduate School of Business studied 193 blockchain initiatives and found that 86% represented material improvements over existing solutions for social impact.[21] Philanthropy was among the most active sectors. But here's what the researchers missed: they were studying isolated experiments. Individual organizations building individual systems. What's needed is greater collaboration and a combinatorial impact; a Global Charity Commons where every charitable act contributes to shared infrastructure.

Think about what this means practically. Right now, when you donate to a charity, you get a receipt. And maybe an email six months later with a general update. But you have no real visibility into impact. You trust—or you don't. With cairn-chain technology, every transaction is recorded immutably. Your $50 for water pumps in Honduras? You can trace it from your account to the implementing organization to the specific community to the verified outcome. Not because someone's telling you it happened. Because it's etched permanently in the chain.

New Rails Mean New Scaling

Here's where it gets interesting for the sector.

When the railroad came to America, it didn't just move goods faster. It changed what could be built, where people could live, what businesses became possible. The rails created the scale.

---

21   Doug Galen et al., "Blockchain for Social Impact: Moving Beyond the Hype," Stanford Graduate School of Business Center for Social Innovation, 2018. https://www.gsb.stanford.edu/faculty-research/publications/blockchain-social-impact

We're at that moment in philanthropy. The existing infrastructure—separate CRMs for every nonprofit, fragmented data systems, duplicated back-office functions across 1.5 million organizations—can't support what's coming. It's like trying to run Amazon logistics on country roads.

The Global Charity Commons provides new rails. Shared back-office infrastructure. Common data protocols. Interoperable systems that let a small community foundation in rural Ohio connect seamlessly with a grassroots Inuit organization in the Northwest Territory. Not because some tech giant owns both, but because they're building on the same open, permanent infrastructure.

What runs on these rails? Everything changes.

Grant reporting becomes pulling verified data from the chain—not recreating documentation from scratch for every funder. Due diligence? The provenance is already there. Audit trails are built in. Donor acknowledgment and stewardship can be automated because intent and fulfillment are already linked in permanent records.

The overhead ratio debate dissolves. When infrastructure is shared, costs are shared. More mission delivery per dollar. That's genuine efficiency from collective investment in common rails.

The Startup Problem for Legacy Nonprofits

This creates an uncomfortable truth for established organizations.

We're about to see a wave of what I call "charity auto-start" ventures—nimble, AI-native organizations that build from day one on this new infrastructure. They won't carry the legacy costs of outdated systems. They won't have decades of accumulated data locked in proprietary silos. They'll be born on the rails.

Legacy nonprofits face a choice: embrace Empower, Innovate, Collaborate—or watch newer entrants scale past them on infrastructure they refused to adopt.

This isn't hypothetical. I've watched it happen in every sector that faced digital transformation. The organizations that treated new infrastructure as a threat rather than an opportunity? They're the ones we don't talk about anymore.

For established nonprofits, the path forward requires genuine collaboration—not the performative kind where you show up at convenings and nod at shared goals, but the operational kind where you actually integrate systems, share data, and build on common platforms. That's harder. It requires boards and executive teams to let go of some control. But the alternative is irrelevance delivered by competitors who were never burdened with the choice.

Innovation Labs and the Attribution Problem

Now here's where my mind really starts racing.

What happens when we combine human creativity with agentic AI systems in deliberate innovation labs—think-and-do sprints where the goal is solving specific community challenges? We're going to generate an enormous amount of intellectual value. New approaches to food deserts. Novel models for elder care. Breakthrough strategies for youth development.

Who owns that?

In the current system, nobody and everybody—which means the ideas either get hoarded by whoever patents fastest or they disappear into the commons with no attribution at all. Neither outcome serves the sector.

The cairn chain changes this. Every contribution to a collaborative innovation process gets recorded. Not as intellectual property in the traditional sense but as permanent attribution. When a small neighborhood association in Detroit contributes a breakthrough insight that later gets scaled across three continents, that contribution is visible. Verified. Permanent.

This matters especially in contexts where traditional intellectual property protections don't function. In regions where patents are routinely ignored or governments don't enforce copyright, the cairn chain provides an alternative accountability mechanism. You can't steal attribution that's immutably recorded across a decentralized network. The small innovator's contribution doesn't get swallowed by the large actor's marketing budget because the record exists independently of any single institution's willingness to honor it.

Small successes compound. Large successes get traced back to their origins. This can be about fair pay, thanks to tokenized attribution enabling that. It's also about dignity and recognition. A system where contribution at every scale gets acknowledged, not just the contributions that come with press releases and donor walls.

From Trust to Verification

"Trust us" worked for a long time in philanthropy. Donors gave. Organizations reported. The gap between what was claimed and what was verified stayed politely unexamined.

That era is ending.

In a world of deepfakes and AI-generated content, "trust us" isn't just insufficient—it's suspicious. Why wouldn't you verify? What are you hiding?

The Stanford Social Innovation Review's 2019 cover story documented how blockchain was transforming philanthropy in China—a context where institutional trust has historically been low.[22] The technology enabled social entrepreneurs to demonstrate transparency in ways traditional reporting never could. Not because they said they were transparent. Because the record was public, immutable, and independently verifiable.

---

22 Xiaofeng Wang and Kevin C. Desouza, "China's New Model of Blockchain-Driven Philanthropy," *Stanford Social Innovation Review* 17, no. 3 (Summer 2019): 26–31. https://ssir.org/articles/entry/chinas_new_model_of_blockchain_driven_philanthropy

We need that everywhere.

The cairn chain isn't about distrust. It's about moving beyond trust to verification. Every participant in the philanthropic ecosystem—donor, organization, volunteer, beneficiary—can see the same record. Disagreements become resolvable. Claims become checkable. And the organizations doing genuine good work have nothing to fear and everything to gain from systems that prove rather than merely assert.

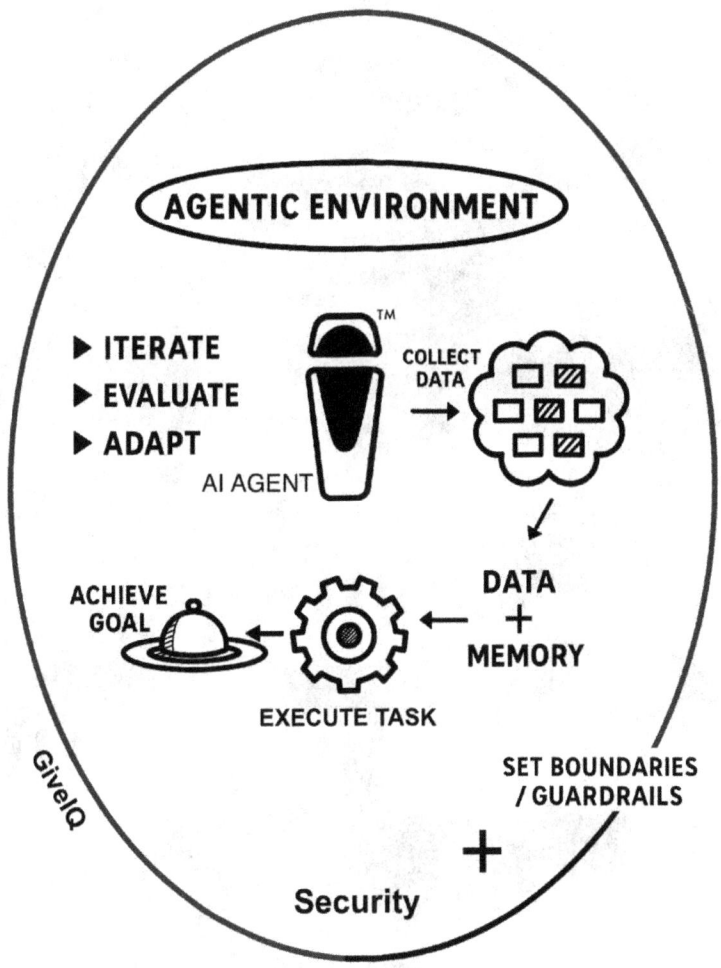

09

THE GLOBAL
CHARITY COMMONS

The Global Charity Commons (GCC)

The Global Charity Commons is a vibrant, borderless, transparent, effective, and efficient philanthropic ecosystem. It represents the full realization of what becomes possible when GiveIQ principles scale globally.

What the GCC Is:

The Global Charity Commons is not a platform, a nonprofit, or a technology stack. It's an intelligent ecosystem, a living network where:

- ❑ **People willing to help** find pathways to meaningful contribution (Charity Autonomy)
- ❑ **Innovation and creative problem-solving** flourish through bounded experiments (Agentic Islands)
- ❑ **Philanthropy large and small** connects, coordinates, and compounds impact

Underpinned by technology like AI and Web 3.0 (soon Web 5.0), the GCC provides the connective tissue that transforms isolated good intentions into coordinated collective action. Smart contracts enable transparent fund flows. AI matches capacity to need. Blockchain creates trustless verification of impact. But the technology is infrastructure, not the point.

The point is what happens when barriers fall: when a grandmother in Boston contributing three hours a week can see her work connect to a youth program in Nairobi; when a software developer's weekend project in São Paulo becomes a template adopted by organizations on five continents; when a small family foundation's $10,000 grant combines with a thousand $10 donations to fund something none could achieve alone.

The Big Pie Philosophy:

Traditional philanthropy often operates with a scarcity mindset. Organizations compete for the same donors, guard their intellectual property, and treat other nonprofits as rivals rather than allies. The Global Charity Commons operates on a different premise: **the pie can grow**.

When participants share best practices, everybody improves. When organizations coordinate instead of duplicate, resources stretch further. When impact data flows openly, learning compounds across the ecosystem. When someone discovers what works, that knowledge benefits every cause, not just their own.

This isn't naive idealism. It's rational strategy in a world where the problems we face, climate change, poverty, health inequity, are too large for any single organization, funder, or government to solve alone. The only path to 100% mission success runs through radical collaboration.

GCC participants and partners commit to:

- ❑ **Transparency:** Open sharing of methods, outcomes, and lessons learned
- ❑ **Interoperability:** Using shared standards so systems can connect
- ❑ **Contribution:** Adding to the commons at least as much as extracting from it
- ❑ **Abundance thinking:** Celebrating others' success as ecosystem success

How the GCC Works:

For Individuals: Your GiveIQ Score and Purpose Path become visible (to the degree you choose) within the commons. You can discover others walking similar trails, find Agentic Islands that match your capacity, and see how your individual contribution connects to collective impact.

For Nonprofits: The GCC provides shared infrastructure, AI tools, data standards, coordination protocols, that would be too expensive for any single organization to build. You gain access to a global talent pool of people seeking meaningful contribution. You can find collaboration partners, share learnings, and benchmark your impact against similar organizations.

For Funders: The GCC offers unprecedented visibility into where philanthropic capacity exists and where gaps remain. You can fund not just organizations but ecosystem infrastructure. You can see how your giving combines with others to move the needle on causes that matter.

For AI Agents: The GCC provides the data, APIs, and governance frameworks that allow agentic philanthropy to operate responsibly, matching contributors to opportunities, optimizing resource allocation, surfacing insights, all while keeping humans in control.

What Makes the GCC Different:

Traditional Model Global Charity Commons
Today, Organizations compete for donors, versus an Ecosystem growing the pie for all. Today, Impact data (intelligence) hoarded versus Impact data shared as a collective competitive advantage resource

Today, Coordination requires formal and slow to act agreements versus Coordination that happens fluidly through partnerships and shared infrastructure

Today, Technology built separately versus Technology pooled and shared by each organization.

SUCCESS= MY ORGANIZATION GROWS SUCCESS = THE CAUSE ADVANCES

Today's Federated Model:

I've seen firsthand the power and limitations of the federated nonprofit model. My first job out of college was with the Boy Scouts of America, with a national membership of then over 4 million youth and hundreds of regional Scouting offices. These offices are called "Councils" and work to grow Scouting in their respective geographic territories.

At Habitat for Humanity, I experienced a similar federated model. Habitat founder Millard Fuller and I met several times during my seven years as the Greater Boston director. From its humble beginning being incorporated in a chicken coop, Millard and Linda Fuller built a respected international organization.

These federated models work, but they're limited by organizational boundaries. The Global Charity Commons suggests a modern alternative. It isn't a single entity but a network that connects people, a digital, interoperable space governed by the principles of Empower, Innovate, Collaborate. It transcends organizational affiliation, connecting anyone willing to contribute with anyone who needs capacity, regardless of which nonprofit's letterhead they operate under.

When Purpose Becomes Collective: The Jones Family

Two months after Gus and her dad each completed their first Agentic Islands, something unexpected happened at their kitchen table.

Gus's mom, who'd been watching both of them quietly, announced she was starting her own island. She'd spent fifteen years as a school nurse but had been thinking about the mental health crisis among teenagers. "If you two can do this," she said, "I can too."

Within a week, Phay.ai had connected her with a local nonprofit piloting peer mental health support programs in schools. Her island: design and test a training curriculum for student mental health ambassadors. Twelve weeks, clear deliverable, built on her nursing experience but going somewhere new.

That's when Gus realized something. This wasn't just about individual Purpose Paths anymore. Her dad helping immigrant entrepreneurs. Her doing digital equity work. Her mom addressing teen mental health. These weren't isolated efforts, they were all responding to real needs in their actual community.

"Phay," Gus asked, "how many other people in our area are working on education and mental health issues right now?"

The answer surprised her. Seventeen active Agentic Islands in their zip code alone, all focused on youth wellbeing from different angles. A retired teacher doing literacy tutoring. A software developer building free college application tools. Three college students running after-school programs. A psychologist offering sliding-scale counseling.

None of them knew about each other. They were all walking parallel trails, solving adjacent problems, but completely disconnected.

"What if they could see each other?" Gus wondered aloud.

The Coordination Problem

This is where individual Purpose Paths start revealing something larger, what's possible when you can actually see the collective pattern of what people care about and where they're contributing.

In my time at various nonprofits, one of the hardest problems was always coordination. You'd have five organizations in the same city all trying to address homelessness, duplicating effort, competing for the same donors, never quite connecting their work effectively. Not because anyone was territorial, but because there was no good way to see who was doing what.

What if every person working on homelessness, paid staff, volunteers, activists, donors actively engaged in projects, had a visible trail record. You could see not just organizations but actual human capacity. Who has housing policy expertise? Who's good at direct service? Who can navigate bureaucracy? Who has relationships with city officials?

That's not a nonprofit database. That's a live map of collective capability.

Gus showed her mom the list of seventeen other people working on youth wellbeing in their area. "You should meet these people," she said.

Her mom looked at the list, a mix of retirees, working professionals, students, people from completely different backgrounds all pulled toward the same general purpose. "How would that even work?"

"I don't know," Gus admitted. "But you're all trying to solve parts of the same problem. What if you could actually coordinate instead of just hoping your individual efforts add up to something?"

The Global Charity Commons Starts Local

This is the vision behind what we're calling the Global Charity Commons, though it usually starts very locally. Not a single platform or organization, but a shared environment where people can discover who else is walking similar trails, what they're learning, where they might coordinate effort.

When Gus's mom reached out to three of the people on that list, the retired teacher, one of the college students, and the psychologist, they met for coffee. Within thirty minutes they'd identified overlaps nobody had seen before.

The teacher was tutoring kids who were struggling academically, but many of them had underlying anxiety or depression affecting their learning. The psychologist was trying to reach teenagers who needed help but wouldn't come to therapy. The local doctor who had been helping youth with ADHD. The college student's after-school program had kids showing up primarily because they had nowhere else safe to go.

"We're all seeing the same kids from different angles," the psychologist said. "What if we actually designed something together instead of each doing our separate thing?"

Six weeks later, they'd launched a coordinated Agentic Island, a pilot program combining tutoring, mental health screening, and after-school support in one middle school. Gus's mom trained student ambassadors who could recognize when peers needed help. The teacher provided academic support. The psychologist supervised the mental health components. The college student managed logistics.

None of them quit their day jobs or previous commitments. They just aligned ninety days of effort around a specific, bounded experiment to see what became possible when their individual capacities combined.

That's not revolutionary. But it's also not how charity usually works. Usually you'd need to form a 501(c)(3), hire staff, write grants, build organizational infrastructure before you could even attempt coordination like this.

The GiveIQ ecosystem makes it possible to coordinate first, formalize later, or never, if temporary coordination is all that's needed.

From Individual to Collective Purpose

What Gus was watching unfold in her own family and neighborhood was a pattern that could scale: people discovering their individual purpose paths, building trail records that prove their capacity, then finding others walking parallel paths and choosing to coordinate.

This changes how we think about social change. The traditional model assumed you needed big institutions, well-funded nonprofits, government agencies, major foundations, to tackle serious problems. Individual effort was nice but ultimately marginal.

But when you can map collective purpose and capacity in real-time, when you can coordinate distributed effort effectively, when you can prove impact through documented trail records, suddenly individual purpose paths can compound into something substantial.

This isn't replacing big institutions. It's adding a layer of coordinated individual agency that hasn't been possible before. The school Gus's mom was working with still existed. The mental health nonprofit still provided infrastructure. But the actual innovation, the pilot program testing a new integrated approach, came from four people coordinating ninety days of effort, supported by AI tools that handled logistics, communication, and documentation.

One Person's Failure, Another's Success

In early 2005, I had the chance to win a State request for proposals bid for an inner city site. My idea was to create the nation's first solely hydrogen-focused new energy incubator. But I was early on the renewables front, especially around hydrogen, and ultimately failed to get any traction. A congressman I met with at the time said, "John, you're on to something big, but you're too early."

Seeing little traction, I tried to pivot by encouraging a nonprofit to take over the high-profile site. The site was adjacent to a highway and, seen by thousands of passing cars each week. I had visited the local food bank that I highly admired and seen their nearby cramped quarters. I used the opportunity to suggest to the CEO that she take advantage of my failure and consider the high-profile site for her own organization's mission impact.

Thankfully, on the second attempt she chose to pursue the idea. Thanks to her and her team's execution, including the prodigious lift to raise the funds and get the site permitted and built, the rest is history, as they say. The food bank has a new high-visibility location that's making a big impact around food insecurity and other regional food needs.

This showcases how one organization's or person's failure can be another's success. In the Global Charity Commons, this can be a regular occurrence. Maybe a donor isn't a fit for one children's charity but can be immediately aligned and connected with a children's hospital working on a cure for pediatric cancers. What a win!

This constancy of connection and network in the Global Charity Commons makes failures in one spot a success in another. That's where agentic philanthropy can be game-changing. The back bench of "outsiders" and underutilized resources doesn't have to stay on the back bench. In a connected commons, someone's abandoned project becomes someone else's launchpad. Someone's mismatched donor becomes someone else's perfect partner. The ecosystem captures value that would otherwise be lost.

The Cultural Shift Underneath

There's something else happening here that's harder to quantify but impossible to miss: People are hungry for this kind of engagement.

Gus's dad didn't just need a paycheck after his layoff. He needed to contribute meaningfully. To use what he was good at. To feel like his twenty years of experience mattered.

Gus's mom didn't need another administrative task. She needed to work on something that addressed what she saw every day in the nurse's office, kids suffering silently with mental health struggles nobody was adequately addressing.

And Gus herself? She could have spent her early years the way most do, school, friends, maybe a low-stakes volunteer gig to pad college applications. Instead she was building genuine capacity, learning what kind of work energized her, developing skills that would matter regardless of which career path she eventually chose.

This isn't altruism. It's something more fundamental, people wanting their lives to mean something beyond consumption and credential-collecting. Values of purpose and contribution, recognized by through history by people like Lafayette.

The old model said: Get educated, get a job, make money, maybe give some back later if you're successful enough. Purpose was deferred, something for retirement or for people who could afford it.

The new model says: Build purpose and capability simultaneously from the beginning. Your trail record compounds over time. Economic opportunity and meaningful contribution aren't separate tracks, they're integrated.

When Gus's dad helps immigrant entrepreneurs, he's not choosing between paying work and volunteer work. He's building documented capacity that creates economic opportunity *because* it's purpose-driven. The community college teaching gig, the consulting work, the board position, all emerged from his trail record of contribution.

AI as Infrastructure for Coordinated Autonomy

None of this requires AI. People have always been capable of coordinating around shared purpose. But AI makes it vastly easier to see who's working on what, connect people who should know each other, coordinate effort across distance and time zones, and document impact in ways that build credible trail records. It again connects back to those neuromorphic ideas on how the brain works. We are the interconnected ecosystem and agentic philanthropy amplifies our connections to do good works.

Phay.ai didn't create Gus's family's Purpose Paths. But it made them visible to each other and to others working on related problems. It reduced the friction of coordination from months of relationship-building to a few coffee meetings. It helped them design bounded experiments instead of open-ended commitments.

This is AI as infrastructure for coordinated autonomy, people acting independently but able to align when alignment creates more impact.

The choice isn't between AI and human agency. It's about whether AI amplifies our ability to act on what we care about, or subtly redirects us toward what serves other interests.

Gus and her family are choosing amplification. They're using tools to walk their Purpose Paths more effectively, coordinate with others walking similar paths, and build trail records that prove their capacity to contribute.

That's not the only way this era could unfold. But it's a way. And it's worth fighting for.

Experiential Governance from Web 3.0 to Web 5.0

Web 3.0 technologies are redefining how nonprofit organizations can engage their communities. This model, Experiential Governance, prioritizes active, immersive participation in organizational decision-making. Experiential governance is not native to Web 3.0 alone, but truly emerges in Web 5.0. Web 5.0 enables purpose and experience, including with authority emerging from contribution of the four T's.

Key Elements:

1. **Tokenized Voting:** Governance tokens grant voting rights on key initiatives
2. **Transparent On-Chain Decision Trails:** Blockchain provides tamper-proof records
3. **Experiential Onboarding via AR/VR:** Donors can walk through a project before voting on it
4. **Mission-Based DAOs:** Stakeholders earn decision-making weight through contributions

Case Study:
Gitcoin Grants, The Global Charity Commons in Action

Gitcoin[23] offers a glimpse of what the Global Charity Commons could become at scale. Using a mechanism called quadratic funding, where the number of contributors

---

23   Gitcoin, "Gitcoin Grants: Funding Open Source and Public Goods," Gitcoin, 2024. https://www. gitcoin.co/grants; See also: Vitalik Buterin, Zoë Hitzig, and E. Glen Weyl, "A Flexible Design for Funding Public Goods," *Management Science* 65, no. 11 (2019): 5171–5187.

matters more than the size of contributions, Gitcoin has distributed tens of millions of dollars to open-source software and climate projects since 2017.

Here's how it works: The community proposes ideas. Contributors signal their support with small donations. The funding algorithm then amplifies projects with broad community backing, ensuring that a project supported by 1,000 people giving $1 each receives more matching funds than a project supported by one person giving $1,000. This mathematically encodes the democratic principle that widespread support should carry more weight than concentrated wealth.

Every action is recorded on-chain, transparent, immutable, auditable. No intermediary decides which projects are "worthy." No gatekeeper extracts fees for access. The community itself becomes the curator, the funder, and the accountability mechanism.

What makes Gitcoin a prototype tool for the Global Charity Commons isn't just the technology, it's the philosophy. Traditional philanthropy asks: "How do we get wealthy donors to fund important causes?" Gitcoin asks a different question: "How do we aggregate the judgment and resources of thousands of people who care?"

The results have been remarkable. Climate initiatives, educational tools, public health projects, all funded not by a foundation's program officers but by the collective wisdom of engaged communities. When the community gets it wrong (and they sometimes do), the feedback is immediate and visible. When they get it right, the model spreads.

Now imagine this mechanism applied not just to open-source software but to every charitable cause on earth. Imagine local communities in Lagos, Lima, and Louisville all using the same infrastructure to fund local priorities while contributing to global coordination. Imagine your GiveIQ score reflecting not just what you gave, but how your giving combined with others to move the needle on problems you care about.

That's the promise of the Global Charity Commons. Gitcoin proves it's not utopian, it's operational.

The Human Capital Engine

At its heart, the Global Charity Commons is a human capital engine. It's designed to unlock the latent philanthropic potential in billions of people who want to contribute but lack the infrastructure, information, or coordination to do so effectively.

Traditional philanthropy has always been constrained by transaction costs. It takes time and money to identify worthy causes, verify their legitimacy, coordinate contributions, and track impact. For large donors with staff and advisors, these costs are manageable. For ordinary people who might give a few hundred dollars a year, the friction often isn't worth the effort.

The GCC changes that equation. By creating shared infrastructure, common data standards, verified impact metrics, transparent fund flows, AI-powered matching, it dramatically reduces the cost of participating meaningfully in charitable work.

This isn't just about making giving easier. It's about making giving smarter. When a mother in Manila can see exactly how her contribution combines with thousands of others to fund a specific classroom in her neighborhood, and can track the educational outcomes of the students in that classroom over time, giving becomes a different experience.

Ask Yourself

For Individuals:

1. What skills do I need to turn my tech expertise into purpose-driven impact?
2. How can agentic and decentralized tools challenge old habits in nonprofits?

For Charities:

1. Are we prepared to bridge the AI skills gap?
2. What would it look like if our tech solutions were built with the community, not just for it?

The Global Charity Commons is where human agency and agentic philanthropy finally meet at scale. It's where the equation, human agency multiplied by agentic philanthropy equals greater societal impact, stops being a formula and starts being a lived reality for millions.

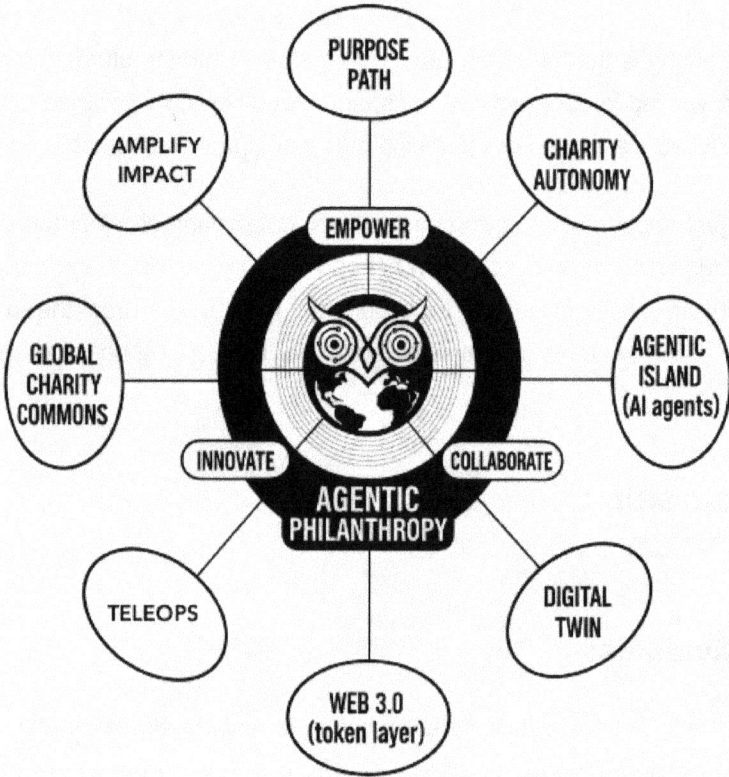

GiveIQ™

THE GIVEIQ
EQUATION

🔑 Key Takeaways

☑ In the age of AI, human agency becomes the scarce resource—not money, not information, not coordination

☑ EIC represents three human agencies: Empower (the agency to act), Innovate (the agency to create), Collaborate (the agency to connect)

☑ The Four T's—Time, Talent, Treasure, Trust—are inputs. What people contribute. Not measures of virtue

☑ Agentic philanthropy multiplies human agency. It doesn't replace it

☑ The 100% Commitment: organizations adopting GiveIQ commit to 100% improvement in mission impact within five years

The Evolution of Stewardship

A charity's board of directors is their chief governance body. This is the group that should be working with the executive director on strategy, securing resources, and serving as community ambassadors for the mission.

I've seen great boards. I've seen not-so-great boards. And I've seen boards stuck in a stasis pattern.

The signs of stasis are easy to spot. Almost no turnover in board seats—folks serving too many years. Low participation and discussion at meetings. They don't prioritize the charity in their annual giving. They're not connectors. They do little to set up a meeting with a prospective donor or a possible partner.

The board exists. But it's not really working.

Meanwhile, AI systems are accelerating stewardship in ways that would have seemed like science fiction a decade ago.

There's a Colorado-based company I spoke with that uses machine learning on satellite imagery. They work with national insurance carriers to determine who gets coverage and at what rates—based on the condition of their properties. They told me something remarkable: they can often determine, by how well a person cares for their lawn, whether the household is too much of a risk to offer insurance coverage.

Think about that. An algorithm analyzing satellite images of your yard to assess your reliability as a steward of property.

Or consider Syngenta's TomatoVision—automated greenhouses using AI and robotics to optimize tomato breeding and environmental controls to improve yields. The system monitors, adjusts, learns, and improves. Constantly.

AI as the overseer of stewardship is here.

The intelligence infrastructure is doubling. Millions more GPUs come online each week. The capacity for monitoring, analysis, and optimization is expanding exponentially.

So one might ask: how does tomato breeding connect to a board of directors?

It's about stewardship. It's about a new kind of amplified accountability.

Why Coherence Matters Now

We've covered a lot of ground in this book. Charity Autonomy. Agentic Islands. The Global Charity Commons. The 100% improvement goal.

These aren't separate initiatives. They're facets of a single insight.

The insight: in the age of AI, human agency becomes the scarce resource.

Not money. Not information. Not even coordination.

Agency. The capacity to choose, to act with intention, to shape outcomes rather than be shaped by them. And this agency needs to be kept sacred in a world where we will increasing look to technology to make things easier for us.

Everything in this book flows from that premise. Every framework, every tool, every recommendation exists to protect and multiply human agency in work of purpose and philanthropy.

That's the equation at the heart of GiveIQ.

As an executive director, I found my best board members were the ones who would reach out to me. Just pick up a phone and ask questions. Maybe it was a point of clarification. Maybe it was an offer to help. The connection mattered.

The worst ones would let things stew. They might reach out to staff members they had known for years and talk about their own confusion. As one might guess, this didn't help the overall culture of transparency and trust.

The evolution of stewardship is a gamechanger for how charities improve. And in parallel, it's how individuals hike their Purpose Paths and achieve their EIC objectives.

EIC: Three Human Agencies

The three pillars of GiveIQ—Empower, Innovate, Collaborate—aren't organizational values. They're descriptions of human agency in action.

Empower is the agency to act. It's your capacity to show up, to do the work, to make the phone call, to write the check, to mentor the young person. Before you can contribute meaningfully to any cause, you need the power to act. Some people have this power constrained by circumstance. Others have it and don't use it. Empowerment means building your own capacity and helping others build theirs.

Innovate is the agency to create. It's your capacity to see what isn't working and imagine something better. To try approaches no one has tried. To bring your full creativity to problems that resist old solutions. Innovation doesn't require inventing something unprecedented. It requires refusing to accept that the way things are is the way they must remain.

Collaborate is the agency to connect. It's your capacity to work with others, to combine efforts, to build something no individual could build alone. The problems worth solving are too complex for solo actors. Collaboration means your impact doesn't end with you. It ripples outward through relationships and networks.

These three agencies are universal. Everyone has them, in some measure. The question is whether you're exercising them, and whether the systems around you are designed to amplify them or suppress them.

A traditional nonprofit model often suppresses agency. Donors write checks and hope for the best. Volunteers do assigned tasks without understanding the larger strategy. Board members attend meetings but don't contribute their full expertise.

The agency is there. It's just not activated. And it's not activated because many times the nonprofits don't have the resources to mobilize that agency. And nor do the volunteers know where their agency is needed most.

The GiveIQ approach does the opposite. It creates structures that activate agency at every level.

The Four T's: What People Contribute

Time, Talent, Treasure, Trust. These are the four ways people contribute to philanthropic work.

They're not virtues to cultivate or metrics to maximize. They're inputs. Resources that flow from people who care toward causes that matter. They are also areas where each one of us can improve by doing. Learning more, and by gaining experience.

Time is your hours and attention. The minutes you spend volunteering at the food bank. The afternoon you dedicate to mentoring. The evening you give to a board meeting.

Talent is your skills and expertise. The marketing knowledge you bring to a nonprofit's campaign. The legal expertise you contribute pro bono. The technical skills you apply to a charity's database.

Treasure is your financial resources. The donation you make. The matching gift you facilitate. The investment in a cause you believe in.

Trust is your relationships. The friend you invite to volunteer. The colleague you connect to an organization. The network effects you create by advocating for a cause.

Traditional philanthropy overweights Treasure. The size of your check defines your status as a philanthropist. But the other T's matter just as much. Often more.

A nonprofit doesn't just need money. It needs hours from dedicated volunteers, expertise from skilled professionals, and network connections from people who can open doors. An organization flush with Treasure but starved of Time, Talent, and Trust is fragile.

The GiveIQ matrix—EIC times Four T's—maps twelve distinct ways to contribute. Not to create a scorecard for judgment, but to reveal opportunities.

Where is your agency flowing? Where is it blocked? What channels remain unopened?

The Equation

This is the equation:

Human Agency x Agentic Philanthropy = Greater Societal Impact

Notice what comes first. Human agency. Your irreplaceable capacity to care, to choose, to show up.

AI doesn't drive the philanthropic enterprise. You do. AI multiplies your effort.

That's not just rhetoric. It's architecture.

Every tool described in this book is designed with the human at the center. Phay.ai doesn't tell you what to care about. It helps you clarify what you already care about and find pathways to act on it. The Global Charity Commons doesn't direct your contributions. It makes the landscape visible so you can navigate it intentionally.

The danger with powerful tools is that they become masters instead of servants. This is the feudal AI scenario I described earlier. Systems that harvest human generosity instead of amplifying human agency.

GiveIQ exists to prevent that.

To ensure that as AI becomes more capable, humans remain more capable too. Not less relevant. More powerful.

Stewardship Transformed

Let's bring this back to boards. How does the equation change stewardship?

Take a group of board members who come to quarterly meetings for the fellowship. Nothing wrong with that, as long as they're contributing. But how do you know if they're contributing? And how do they know what's expected?

EIC stewardship parameters can now be set up in advance—like the algorithms for the tomato environments we talked about earlier—to make sure board members understand the charity's strategy, their role, and where they might contribute the most.

The executive director of a small music charity in New York City told me she clearly understood the strengths and weaknesses of her small board. She knew who could get her a meeting at a New York public school. She knew who could secure a larger donation from their network of high-net-worth individuals. She knew who she could rely on to come through.

The early introduction of an EIC dashboard provides clarity to both the board member and the executive director. They get on the same page. They understand the contribution expected.

Maybe it's the agency to act, and the board member needs to help identify and deliver a new volunteer to conduct the organization's audit.

Or maybe it's an expectation of writing a check—or finding someone else to write a $5,000 donation.

Or maybe it's making three introductions this quarter to potential corporate partners.

Agentic philanthropy elevates stewardship—including the fiduciary, legal, and strategic responsibilities of a board. This is a gamechanger for an under-resourced nonprofit.

Twelve Islands in One Year

A hypothetical, regional food bank had relied on the same volunteer model for decades: sorting and packing food. They had plenty of sorting volunteers. What they lacked was strategic capacity.

We designed twelve Agentic Islands for their first year:

A graphic designer spent 40 hours creating a nutrition labeling system. A logistics professional devoted 60 hours to transportation route optimization. A researcher conducted a 50-hour partner agency survey. A data analyst performed a 45-hour donor retention analysis. A bilingual contributor spent 55 hours translating all materials into Spanish.

A development professional researched grant prospects for 35 hours. An HR professional developed a board recruitment strategy over 40 hours. A marketing professional created a social media content calendar in 50 hours. An IT professional evaluated volunteer management software for 45 hours. A food safety professional built a training curriculum over 60 hours. A sales professional developed a corporate partnership pitch deck in 35 hours. And a program evaluator created an impact measurement framework over 70 hours.

Eleven of twelve islands were completed within the year.

Total contributed hours: approximately 570. That's equivalent to a quarter of a full-time staff position. But these weren't general hours. They were expert hours applied to strategic challenges.

The food bank's executive director summarized it simply: "We accomplished more strategic work this year than in the previous five years combined. And it didn't cost us anything except learning how to manage differently."

That's the equation in action. Human agency—skilled professionals willing to contribute—multiplied by agentic philanthropy—bounded projects with clear deliverables and structured support. The result: strategic capacity that transformed the organization.

The 100% Commitment

Every organization that adopts the GiveIQ approach commits to a bold goal: 100% improvement in mission impact within five years.

This isn't about working twice as hard or raising twice as much money. It's about working smarter: eliminating waste, unlocking latent capacity, coordinating more effectively, and focusing resources on what actually drives outcomes.

The eradication of Guinea worm disease shows what's possible when organizations fully commit to mission success.

When The Carter Center assumed leadership of the global eradication campaign in 1986, an estimated 3.5 million people were infected annually across 21 countries. By 2024, that number had dropped to 15 cases—a reduction of more than 99.99%. Guinea worm is poised to become the second human disease in history to be eradicated, following smallpox, and the first to be eliminated without a vaccine or medicine.

The entire effort was built on behavior change, community-based interventions, and relentless coordination. As Kelly Callahan, a Carter Center public health worker, explained: "Guinea worm disease has no cure, no vaccination. Basically the entire eradication effort is built on behavior change."

Human agency, coordinated and sustained over decades, producing results that seemed impossible.

GiveIQ argues that this kind of transformative impact shouldn't be rare. It should be the norm.

For some causes, 100% improvement may be a stepping stone to something even more ambitious: 100% mission success. The complete elimination of a problem. The full achievement of a goal. These outcomes become imaginable when human agency is properly activated and amplified.

Simple in Statement, Profound in Implication

Human Agency x Agentic Philanthropy = Greater Societal Impact.

Your agency flows through EIC channels—acting, creating, connecting.

It takes the form of Four T contributions—Time, Talent, Treasure, Trust.

Agentic philanthropy multiplies each step. AI helps you clarify your Purpose Path. It matches your capacity to needs that fit. It coordinates your efforts with others working toward the same goals.

The result is societal impact at scales we've never achieved. Not because the technology is magical.

Because human agency, properly amplified, is more powerful than we knew.

That's the GiveIQ equation.

Simple in statement. Profound in implication. Ready to be tested against the real world.

Ask Yourself

For Individuals:

1. Which of the three agencies—acting, creating, connecting—comes most naturally to you? Which have you neglected?
2. Of the Four T's, which do you contribute most readily? Which remains untapped?
3. If you mapped your philanthropic activity against the twelve dimensions (EIC × Four T's), where would you find the most energy? Where would you find silence?

For Organizations:

1. If a hundred skilled, passionate people showed up tomorrow wanting to contribute, could you deploy them effectively? Or would you become the bottleneck?
2. Do you have a portfolio of bounded projects ready to absorb new capacity—quick wins, standard projects, and deep dives?
3. What would 100% improvement in mission impact look like for your organization? Is it even imaginable with your current model?

MOVING TO THE EDGE: HUMANOIDS FOR BETTER HUMANS

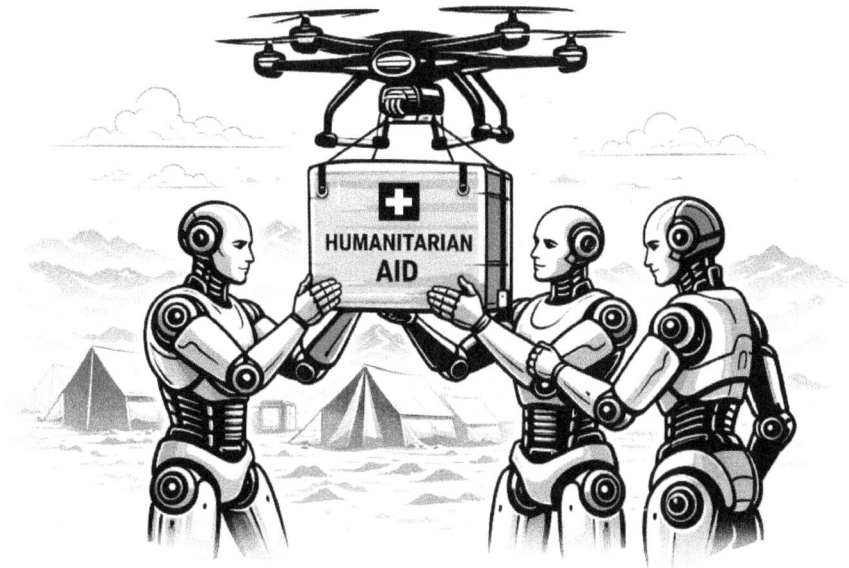

🔑 **Key Takeaways**

☑ **Teleoperation Bridges Distance:** Humanoid robots and teleops enable volunteers to contribute meaningfully from anywhere in the world, bringing human empathy to robotic capability.

☑ **Technology Must Enhance, Not Replace:** As robots become ubiquitous, nonprofits must ensure automation enhances human potential rather than displacing vulnerable workers who find meaning and dignity in their work.

☑ **Senior Care Revolution:** AI companions and robotic assistants can combat the epidemic of loneliness among seniors while preserving human connection and dignity—but only if we design them to connect people to more humans, not fewer.

Augustina Jones and the Rise of the Humanoids

"Five, four, three, two, one, connection secure. You're a go, Gus." The countdown readied Gus Jones to take control of the next generation of teleoperation, "Assist-techs." Her headset provided a brain-computer interface (BCI) for Gus to fully immerse herself into the midst of a refugee camp just outside of French Guyana. There had been two weeks' worth of extreme weather events that had led to mudslides, building destruction, and mass migration. Gus had volunteered herself and a team at Cyberguardians to spend a week helping with humanitarian assistance.

The latest generation of AI-underpinned tools were married to teleoperations and humanoid robots. Assist-techs were exoskeletons that enabled a volunteer to participate from anywhere in the world. Through experiential training, one can quickly learn how to operate the Assist-techs hybrid of teleoperations and humanoid robotics. Humans would provide the vision and empathy, while the robotic components would do the heavy lifting and coordinate with AI agents. For the Global Humanitarian Assistance League (GHAL), volunteers adept at using Assist-tech were invaluable.

Extreme weather events have accelerated worldwide, and consequently, destruction has occurred along with it. The most vulnerable communities tended to be both poor and urban. And with most of the world living in urban areas, the work of GHAL was in high demand. GHAL galvanized thousands of people and AI to restore communities in more climate-adaptive ways. GHAL's volunteer corps augmented paid staff by 100-fold—for every GHAL staff member, 100 volunteers were helping out. On top of that, thousands of AI agents supported those respective teams.

Gus took a deep breath, adjusting her AR headset as the real-time data flooded her field of view. The refugee camp was a sea of movement. Drone bots worked in unison with families to set up shelters and with aid workers distributing food. The destruction was vast, but so was the determination to rebuild.

"Gus, we need Assist-techs over at Zone B," came the voice of Anika, one of the on-ground coordinators for GHAL. "We've got debris removal underway for the 3D housing units, but the site's still unstable." A digital twin appeared in Gus' augmented reality dashboard, complete with immediate action steps that needed to be taken to shore up the unstable ground.

"I'm on it," Gus responded, guiding her Assist-tech toward the rubble-strewn area where robotic arms and AI-driven cranes were already clearing the remains of collapsed structures. Working with her was a next-generation group of landscape architectural students from the Rhode Island School of Design (RISD) Resilience Builders Program. The students maneuvered their own Assist-techs to join Gus. They all remotely helped excavate to clear a path for the 3D-printed homes taking shape nearby. This volunteer service work provided the students with unmatched experiential learning.

The hum of the 3D home printing machinery filled the air as massive robotic arms extruded layers of sustainable concrete polymer. These new storm-resistant homes were a sight to see. Assist-tech crews would swoop in to complete the construction, including the hook-ups to the local utilities and the fine details inside the homes. Gus marveled at how quickly the urban landscape was changing—just days ago, this was a field of wreckage, but now, walls stood where ruin had once been.

Above them, a fleet of construction drones hovered, carrying heavy bundles of native vegetation to shore up the slope that had been devastated by the mudslide. AI-assisted designers prioritized the terrace design that took shape to secure the slope's soil better. Automated soil sensors guided the drones, ensuring the plants took root in the right spots to construct an almost lattice architecture of sustainable landscaping.

"Gus, heads up—drone drop incoming," one of the RISD students called out—alerted by their personal AI construction agents. Gus shifted her Assist-tech to grab a bundle of saplings as a drone released them midair. She and the students planted in the AI-designed lattice pattern, which benefited from data and best practice sharing from other sites around the world.

As the sun dipped lower, a new skyline was forming, not of glass and steel but of resilience and renewal. These homes, reinforced with climate-adaptive materials, weren't just shelters; they were symbols of a future where disasters didn't just destroy; they catalyzed global empowerment, innovation, and collaboration.

Gus wiped the sweat from her brow. Despite being thousands of miles away from the actual construction zone, she saw, heard, and felt everything around her. She stood back for a moment to admire their progress. She was comforted with the knowledge that her Assist-tech would soon be seamlessly transferred to another dedicated volunteer—who would pick up where Gus left off. GHAL's teleoperations created a perfect continuity to accelerate an around-the-clock construction schedule.

She thanked her RISD co-workers and brought her Assist-tech exoskeleton back to its recharging base. A new sustainable neighborhood was quickly taking shape, and soon, the laughter of children and families would fill the air.

The Robots Are Here

Growing up, I loved science fiction shows like Star Wars and Star Trek. One of my earliest introductions to the genre was the Lost in Space series, where I was fascinated by the "robot." I still think the Lost in Space robot had such a cool design that sparked many of us' curiosity about robotics. Its transparent glass head resembled an exotic sparkling fish tank, swiveling 360 degrees with colorful lights and antennae moving inside. The robot's shiny metallic torso made it look more formidable, complete with a colorful electronic panel. Its accordion arms ended in pinching red hands, and the robot's commanding voice often gave fatherly advice to the show's lead character, a boy named Will Robinson. "Danger, Will Robinson!" was the robot's iconic warning whenever they faced the dangers of being lost in space and encountering alien adversaries.

Robots are no longer just science fiction; they'll soon reshape our work and lives. Elon Musk's humanoid robot, Optimus, priced around $20,000, aims to perform repetitive tasks, with Musk expecting high sales. Boston Dynamics' Atlas and Spot demonstrate advanced agility and applications from construction to healthcare.

Robots will likely be rented for short projects or long-term work, and home robots like robotic lawnmowers are now available under $1,000. Future lawn robots might also handle weed removal, fertilizing, and clippings. Residential robots with bionic features will soon manage mundane chores, rapidly integrating into daily life.

This pattern of technological change isn't new—printing presses, automobiles, computers, and the internet have all transformed society. As robots become common on streets and in stores, similar to cars replacing horse-drawn buggies, they could be a daily sight by 2035.

At first, household robots will be luxury items. However, in short order, households will stretch to buy robots, and the prices of mass-produced robots will start to fall. The first robots will be like the first mass-produced televisions, which were introduced in the late 1940s. For my grandparents, it was a stretch to buy a television. My grandfather was a fireman, and my grandmother worked in the hat department at Boston's Filene's Department Store. But they budgeted and saved up and became one of the first families on their block in Dorchester to own a TV. The television was a large, wood-panelized box—comparable to a large office desk or a clothes bureau.

The TV took up so much room in one's living room that everything else had to be designed around it. The couch was immediately across from the television. If you had a fireplace, the television had to be away from that heat source. And it couldn't be near a window because of the reflection of light coming in. The "theater-ization" of the home began with the introduction of these giant picture boxes. Regular broadcasts meant families would begin to plan their schedules around their favorite shows. These hour-long shows gave way to hours of shows and were the start of a more sedentary lifestyle. The advent of television also reinforced how we like to connect with stories and learn visually.

The first robots might be hard to miss, like those early televisions that took up so much room in one's home. But eventually they will begin to blend into our everyday. Humanoids, the latest generation of robots, are infused with both AI and mechanical adroitness, bringing our long-held anthropomorphic designs to life. Some scientists call it embodied intelligence. These humanoids will eventually be everywhere and indispensable—a digital companion that possibly, like the family dog, shadows our every move.

As a species, we have always worked to design things with human-like features to help us connect and better relate to our creations. Honda's ASIMO, with its childlike stature and expressive movements, was designed to help people feel more at ease around robots, serving as a guide or assistant in public spaces. Similarly, Pepper, a social robot developed by SoftBank Robotics, was created to interact with humans on an emotional level, understanding and responding to emotions, making it an effective companion in retail and healthcare settings.

What's Next: From Assistants to Amps

AI inference involves AI systems applying learned patterns to predict outcomes with new data. Advances like AI-augmented robots that can build and train other robots could manage warehouses, assemble parts, or use 3D printing autonomously. These robots will integrate into large networks, accessing inventories and customizing components on demand. Soon, drones might deliver parts to your door, with humanoids installing them. This rapid AI evolution raises safety, ethics, and governance issues, requiring safeguards to ensure responsible use aligned with human values.

As AI becomes deeply embedded in robotic systems, some robots will reach a "master's" level of collaboration, orchestrating complex networks of specialized agents. Let's call these advanced robots "Amps" after the electronic devices musicians use to amplify sound. In the robotics world, Amps will serve as powerful hubs, enhancing the capabilities of an entire agentic network.

We are only in the early days of understanding what the agentic layer of AI will look like. Amps, along with their subordinate specialized agents, will operate 24/7, tirelessly executing tasks, learning, and evolving. They are a precursor to Artificial General Intelligence (AGI), exhibiting remarkable intelligence, adaptability, and speed. Much like how your smartphone updates overnight while you sleep, Amps will be constantly refining and executing tasks at all hours. And, like the AI in the movie Her, these embedded AI systems will be able to interact with thousands of AI agents and humans simultaneously.

The foundation model sizes are growing from hundreds of millions of parameters to trillions of parameters. That means they are learning much, much more—learning the languages we speak and our human behaviors. This is called social reinforcement learning (SRL). SRL understands visual cues in daily interactions with humans and thereby helps to continually improve the efficiency and quality of work. Just by looking at us, these robots will soon detect if we are hungry, have a question, or are experiencing an emergency. A transformation will happen from robots that are proficient around the home and workplace to those who evolve into daily companions.

The Slippery Slope of Digital Companionship

There is a slippery slope on companionship though. "I think about AI agents like a digital companion," says Nancy Xu, CEO and founder of MoonHub. Social media has been the canary in the coal mine in terms of how algorithms have gotten to know us. This includes algorithms to motivate us to want "likes" on social media platforms like Facebook, Instagram, and TikTok. One can easily think about the cracks that have already been created in our lives thanks to social media and technology overload.

How might we extrapolate that into the ways that AI and humanoid companions may both augment human activities but also affect our humanity? While we are rushing to have robots take over the mundane and mechanical, we need to ask what is sacred to us and what is uniquely human labor.

Research studies have attributed our overuse of technology to shrinking attention spans and declining cognitive health. A study by Microsoft showed that our attention spans had decreased from 12 seconds in 2000 to less than 8 seconds by 2013. There's even a term called digital dementia, coined by neuroscientist and psychiatrist Manfred Spitzer, that's used to describe memory problems and cognitive decline caused by an overreliance on technology. Our overreliance on the digital or technological world will rob us of even more cognitive abilities as we introduce robots into our lives. Robots will, after all, have IQs that are conservatively several factors higher than our own.

The Role of Teleops in Solving Global Challenges

The integration of humanoids, robotics, and teleoperation (teleops) into philanthropy represents an unprecedented opportunity to address some of the world's most pressing challenges. By enabling remote human control over robotic systems, teleops bridge the gap between technological ingenuity and localized problem-solving, empowering communities and aiding organizations to co-create impactful solutions.

Robot teleop refers to teleoperation, which is the process of remotely controlling a robot. In teleoperation, a human operator directly controls the robot's actions using an input device such as a joystick, game controller, or computer interface. It is often used when autonomy isn't feasible or when precise human guidance is required.

Applications of Robot Teleop

In robotics, competitions like FIRST showcase real-time human-robot interaction, especially during 'teleops' where humans control robots, highlighting teamwork and decision-making. Teleoperation also benefits remote exploration in hazardous or inaccessible environments, like underwater ROVs or space rovers, enabling safe, precise guidance where human presence is risky. In industry, teleoperated robots perform complex manufacturing and construction tasks requiring human finesse, such as welding or handling delicate materials. The medical field benefits most, with systems like Da Vinci enabling surgeons to perform precise, remote procedures, transforming patient care.

Stacking Stones: How Teleops Builds Its Own Trail

The teleops framework serves as a blueprint for deploying robotics and AI in charity work, offering a structured approach to ensure that technology truly serves community needs.

In the future, robotic teleoperations could become as ordinary as flying a hobbyist drone. Larger drones and more capable drone pilots could deliver more complex or urgently needed goods to remote or disaster-affected regions. But drone delivery is just the beginning. As the technology matures, teleoperation will extend to ground-based robots capable of construction, medical assistance, and infrastructure repair—tasks requiring dexterity, judgment, and real-time human guidance. The United Nations and other international bodies can take the lead in creating a global aid infrastructure built around GiveIQ principles, including teleoperations. On-the-ground sensors, encrypted blockchain communications, convenient landing zones, charging capacity, and trained operator teams are the building blocks of a minimally viable product for global aid robotic teleops.

What makes this vision more than science fiction is the learning loop at its core. The data generated through teleoperation will be invaluable for training the next generation of AI systems. These real-world experiences provide detailed data points for reinforcement learning and imitation-based policy development that simulations alone cannot capture. When human operators guide robots through complex tasks—navigating unpredictable environments, manipulating delicate objects, coordinating whole-body movements—they generate rich visual-spatial datasets that encode not just *what* to do, but *how* to adapt when conditions change.

Research published in *The International Journal of Robotics Research* demonstrates that combining teleoperated demonstration data with unsupervised robot interaction enables systems to learn visual dynamics models and action capabilities that transfer to novel, previously unseen scenarios (Ibarz et al., 2021). The researchers found that approximately 500,000 real-world training trials—incorporating visual sensor data from teleoperated demonstrations—achieved generalization capabilities comparable to what ImageNet's million images provide for object classification. High-quality teleoperation data, capturing the nuanced decision-making of human operators in three-dimensional space, creates the foundation for robots to develop genuinely adaptive intelligence rather than merely executing pre-programmed routines.

What makes this data particularly valuable is its embodied nature: operators must perceive depth, anticipate physical forces, and make split-second adjustments based on sensory feedback. Every hour of skilled teleoperation becomes a resource for training systems that can eventually handle more tasks with less direct oversight.

Learning at the Edge

But where should this learning happen? Traditional AI systems aggregate massive datasets in centralized servers, process them, and push updates back to devices. This works for predictable environments. It fails in humanitarian contexts, where connectivity is unreliable, conditions change by the hour, and the data that matters most—a washed-out road, a crowd gathering at a distribution point, a child's outstretched hand—can't wait for a round trip to the cloud.

Neuromorphic computing offers an alternative architecture. Rather than processing every input through dense neural network layers, neuromorphic systems mimic biological neurons: they activate only when sensory input crosses a threshold, they process information locally, and they adapt in real time. This "sparse, event-driven" approach dramatically reduces power consumption and latency—critical advantages when your robot is running on solar power in a refugee camp or operating in a communications-denied environment.

Here's where teleops becomes transformative: every human-guided operation becomes a training signal at the edge.

When a remote operator guides a delivery drone through turbulent crosswinds to reach a flooded village, that sensorimotor data—the precise adjustments, the depth perception, the anticipation of obstacles—doesn't need to be uploaded to a central server for batch processing. With federated learning architectures, the drone's local model can update immediately, sharing only the model parameters (not the raw data) with other drones in the fleet. The learning stays distributed. Privacy is preserved. Bandwidth requirements plummet. And the next drone approaching that same village already knows something about the wind patterns.

Research in organic neuromorphic electronics has demonstrated that decentralized sensorimotor circuits can form associative learning links in real time, enabling robots to navigate novel environments after minimal training (Krauhausen et al., *Science Advances*, 2022). The key insight: when sensing, computing, and acting happen at the same location—rather than being separated by network latency—adaptation becomes immediate. Intelligence gets woven into the fabric of the system itself.

For humanitarian teleops, this suggests a new operational model:

❑ **Edge-first learning:** Rather than treating field robots as data-collection terminals for centralized AI, design systems where learning happens primarily at the edge, with human operators providing real-time correction that updates local models immediately.

❑ **Federated wisdom:** Robots operating across different humanitarian theaters share model updates rather than raw data. A drone learning to navigate monsoon conditions in Bangladesh contributes to the collective intelligence of drones operating in flood zones globally—without any single server holding sensitive location or beneficiary data.

❑ **Neuromorphic efficiency:** Deploy processors that activate only on threshold events, reducing power consumption by orders of magnitude. In environments where charging infrastructure is scarce, this isn't a nice-to-have—it's the difference between operational and useless.

The Human in the Loop

But here's the critical point: *more capable* should not mean *fully autonomous.* As agentic philanthropy develops increasingly sophisticated capabilities, the temptation will be to remove humans from the loop entirely. This would be a mistake. There are some things we'll resign ourselves to ceding—the speed of compute, for instance, is already beyond human scale. But the rise of truly autonomous systems will falter, and should falter, when humans are kept out of the learning loop. Human oversight isn't a limitation; it's a safety architecture.

Keeping humans in the loop is the lower-risk path for prioritizing human safety and control. This is a semi-autonomous approach that scales with human agency, intelligent research, and iteration at its core. Guardrails that keep strategic decisions in human hands—including what a teleops system learns next and which other systems it communicates with—are critical decision points. Think of it like a child checking in with a parent: have fun, explore, but let me know when you're leaving home and heading outside the neighborhood.

The human operator isn't a bottleneck in this model—it's the training signal that keeps the system aligned with human judgment and values. When an operator decides to deviate from an algorithmically optimal path because they see a family waving for help, that decision becomes part of the system's understanding of what humanitarian response actually means. The machines learn not just *how* to navigate, but *why* certain choices matter.

Sensorimotor Stigmergy

Ultimately, these data points and best-practice learnings become the stones for the proverbial cairns I've followed on countless trails. In complex systems theory, this is called *stigmergy*—the phenomenon where the environment itself becomes the communication medium. Ants leave pheromone trails; hikers stack stones; teleops systems leave behind training data and operational patterns that guide what comes next.

I call this approach **sensorimotor stigmergy**: the environment becomes the memory. Each teleoperated mission deposits learning that shapes future missions, creating an emergent intelligence that no single operator or robot possesses, but that the system as a whole can access. The accumulated sensorimotor wisdom becomes the trail markers—the cairns—that guide subsequent operations without explicit communication.

Just as ants leave pheromone trails that guide colony behavior without centralized control, teleops systems leave learned patterns in the operational environment itself. A drone that successfully navigates a displaced persons camp leaves behind model updates that make the next drone's job easier. A robotic arm that learns to sort medical supplies in a field hospital contributes that knowledge to every similar system in the network. No central authority needs to coordinate this. The trail builds itself.

The system self-organizes to keep us on the trail, each action shaping the landscape for those who follow. In agentic philanthropy, the cairns are built from accumulated wisdom—human and machine—marking the path toward more effective, more responsive, more humane aid.

This is neuromorphic philanthropy in action: a giving ecosystem that mimics how brains actually work, where intelligence is distributed, where learning happens at the edge, and where every purposeful action contributes to the collective capacity to do good. The neurons fire. The stones stack. The trail emerges.

What Goodwill Taught Me About Technology

In the late 1990's I had the privilege of working at Morgan Memorial Goodwill Industries. Each morning, pulling up to our Roxbury, Massachusetts headquarters, I was inspired by the organization's mission, but even more inspired by the people we served. For some, Goodwill meant job training, while for others it meant specialized learning programs.

At the heart of its earned revenue operations was Goodwill's giant recycling operation. That business had become synonymous with the act of giving itself. When cleaning out closets, one often heard, "Let's take it to Goodwill." Over 3.8 billion tons of donated clothing, furniture, and other household items are being collected through 150 U.S. Goodwill locations and at 12 international Goodwill centers.

Beyond sustainability, Goodwill's recycling and sorting created meaningful jobs for those recovering from brain injuries or facing mobility issues, boosting their confidence and sense of purpose. I watched people with cognitive disabilities sort donations. They worked slowly. And they needed kind supervision.

But here's what they had that machines don't: they needed to be needed.

The grandmother who came in to sort clothes three days a week wasn't there for the paycheck (minimal) or the efficiency (low). She was there because someone remembered her name, her coworkers invited her to their birthday parties, she felt useful, she belonged somewhere.

When we automate that job away, when we say robots can do it better, faster, cheaper, we're not just eliminating a task. We're eliminating a reason to get up in the morning.

As robots become ubiquitous, they also risk displacing jobs for vulnerable groups. As shown by Goodwill, these roles are more than tasks—they are crucial for social inclusion, rehabilitation, and personal fulfillment. The question isn't whether to use robots. The question is: What do we owe each other that robots can never provide?

Nonprofits, therefore, face a crucial challenge and opportunity. By proactively adopting technology they can shape technological advancement toward enhancing human potential rather than supplanting it. Such partnerships, leveraging corporate resources and innovation, can amplify nonprofit missions without sacrificing essential human touchpoints. Striking this careful balance is paramount; technology must serve to expand—not diminish—the profound impact organizations like Goodwill have cultivated for over a century.

The Coming Disruption

I believe there will be massive short-term disruptions in workplaces due to AI-driven automation, and these impacts will disproportionately harm those engaged in entry-level roles. Don't get me wrong, there will be new jobs that will proliferate the economy as well, but the current job mix will be overturned. And there are many folks talking about a universal basic income or universal basic services to help us through this rough transition. But I believe this will be a sea change like none other we have experienced.

A conversation I recently had with a C-level executive at a Fortune 50 company underscored this reality. Although they asked not to be quoted, the executive revealed their company was eliminating most positions within its customer call centers, traditionally staffed by large numbers of human representatives. When I asked how their new model would function, they described a fundamental shift in customer service delivery: instead of employing human representatives as the primary point of contact, AI-driven virtual agents would handle the majority of initial customer interactions. Select human representatives would remain, but each would now oversee multiple AI agents simultaneously, stepping in only when customer queries surpassed the virtual agent's capabilities or when customers explicitly requested human assistance.

This hub and spoke approach—with the human as the hub and the AI agents as the spokes—will become a common organization.

According to a recent Brookings Institution study, about 36 million American jobs, especially those involving routine cognitive and manual tasks, are highly exposed to AI-driven disruption in the coming years. The study emphasizes the urgency of proactively developing robust strategies for workforce retraining and economic adaptation to mitigate widespread job displacement.

Senior Care: AI's Profound Promise

Senior care is one area where we will witness profound positive disruption driven by AI. One of the toughest moments in my life came when my sisters and I had to move our mother from her cherished home into assisted living memory care. My Mom had lived independently until age 86, navigating life bravely after our father passed a few years prior.

We visited her weekly, taking her out to dinner, movies, church, doing housework or grocery shopping. We called daily. Little gestures meant the world, helping her feel connected and valued. Yet, despite our best efforts, we saw the loneliness creep in during the long hours we couldn't be there. My Mom was very independent though— one of our favorite photos is her at eighty-six, shoveling the driveway in February, in her trademark skirt and sweater.

Eventually, our visits, her interaction with her wonderful neighbors and the work around the house, wasn't enough. She needed 24/7 care, and we couldn't afford round-the-clock humans. So we moved her to memory care. She was mad at me for the first few weeks, but then seemed to enjoy the socialization, the steady meals and the extra care.

But there's a different scenario coming. What if we'd had access to an affordable humanoid assistant? One that could remind Mom to take her pills, make sure she didn't leave the stove on, alert us if she fell, help her get dressed, and kept her company during the long afternoons. Would that have been enough to keep her home? Maybe. But here's what haunts me: Would she have been less lonely, or more?

The Uncanny Valley of Care

A humanoid can respond with empathetic phrases. But it can't actually feel worried about you. And almost-but-not-quite-human care might be lonelier than no care at all.

The brutal economics are this: humanoid caregivers are coming because human caregivers cost $15,000-20,000 per month. A humanoid costs $20,000 once. Unless we design it differently, we're surrendering to economics rather than making progress.

Loneliness among seniors isn't merely emotional discomfort; it's a serious health issue linked to cognitive decline, depression, and diminished physical health. With aging populations globally, combating senior isolation is becoming an urgent challenge.

Humanoid robots with teleoperation capability can act as companions and care assistants. These robots could be remotely operated by caregivers or family members, enabling regular interaction. Unlike video calls, these robots facilitate physical interaction—holding hands, aiding mobility, or doing household tasks. And with generative AI, these robots could offer memory games, work with the senior on physical therapy or be a regular interface for connecting with family and friends.

Companies like Intuition Robotics developed ElliQ, a social robot aimed at reducing elderly loneliness. Early research shows benefits, including better mental health and increased engagement with loved ones. When AGI comes along at the edge, these companions will be powerful helpmates. They will also be embedded with wonderful personalities, enthralling stories, perfectly pitched singing voices and more.

The robot we actually need: What if we built robots that connected isolated seniors to volunteer visitors? That handled logistics so humans could focus on connection? AI in senior care should scale human compassion—connecting people to more humans, not fewer.

The Rise of Brain-Computer Interface

While we read about Gus Jones and the RISD students helping with exoskeletons and their augmented reality teleoperation headsets, we will shortly have more fully immersive interfaces. The rise of brain-computer interface (BCI) technology will not be just a revolution in human capability but can also lead to a transformation in how we give, connect, and contribute to society. At its core, BCI enhances the brain's natural outlets—empathy, creativity, and decision-making—allowing individuals to engage in philanthropy with unprecedented precision and impact. Imagine a world where thought alone can initiate charitable action, seamlessly directing resources, coordinating volunteer efforts, or even optimizing donor impact in real-time.

> *"The most important revolution of our time is the intersection of human biology and technology."*
>
> **RAY KURZWEIL**

The technology is already proving its therapeutic power. In January 2024, Neuralink implanted its first human patient, 29-year-old Noland Arbaugh, who had become quadriplegic after a diving accident. Within weeks, Arbaugh was moving a cursor on a computer screen, playing chess, and browsing the web—all through thought alone. As of late 2025, Neuralink has implanted multiple patients, accumulating thousands of hours of use, with participants reporting they've regained significant autonomy. "I can play chess, browse the web, and use a phone, all just with my brain,"[24] Arbaugh shared, demonstrating how BCIs can restore what injury has taken away.

The trajectory toward consumer neural interfaces is accelerating. Apple has received a patent for AirPods equipped with biosensor technology capable of measuring electroencephalography (EEG) signals and other biomarkers.[25] The patent describes a system where active electrodes would be embedded in the replaceable ear tips and reference electrodes positioned on the stems and housing, allowing the earbuds to detect electrical activity in the brain. The technology employs dynamic electrode selection, where algorithms intelligently choose the best subset of sensors based on factors like impedance, how the device fits in each individual's ear, and ambient conditions. This development signals a significant evolution toward neural communication interfaces, positioning consumer earbuds as potential brain-computer interface devices that could enable thought-based controls, monitor neurological health conditions like epilepsy or sleep disorders, and create more intuitive human-machine interactions. While the patent raises important questions

---

24 "Neuralink - Wikipedia," Wikipedia, accessed December 2025, https://en.wikipedia.org/wiki/Neuralink.

25 Apple Inc., U.S. Patent Application for "Earbud with Biosensor," describing EEG-capable AirPods with dynamic electrode selection, 2024.

about data privacy and the protection of sensitive neural information, it represents a major step toward making neural sensing technology accessible to mainstream consumers.

As BCIs evolve, they'll bridge human intent and action, enabling charity work at the speed of thought. An AI philanthropic assistant in a BCI could analyze crises, match donors, and direct funds with a neural command, removing traditional friction and increasing trust through real-time verification and blockchain. Tech leaders like Musk and Huang see this merging of human cognition and AI as transformative. Musk envisions a "closer merger of biological and digital intelligence," a vision that could expand powerfully into philanthropy. BCIs could allow immersive charitable experiences using AR or gamification, strengthening empathy and lowering barriers to giving.

But we must proceed carefully. The research on BCI's psychological effects is sobering. According to a 2019 scoping review in *BMC Medical Ethics*, while BCIs offer increased independence and autonomy, they can also lead to "impingements on human autonomy, psychological frustration, the creation of dependency, and confusion regarding user self-perception."[26] A qualitative study by F. Gilbert found that patients using BCIs perceived the devices as an extension of their bodies—"materializing into a portion of themselves"—which raises profound questions about identity and what happens when such devices fail or are removed.

A 2024 European Council report on BCIs raised additional concerns: individuals may experience "cognitive fatigue, skin irritation, headaches or eye strain from extended use," along with a "psychological dimension" that includes the potential for BCIs to "influence emotions, personality and memory." The report noted that the brain's natural mechanism to suppress traumatic memories could be compromised. Perhaps most alarming, the long-term cognitive effects remain largely unknown— only 1% of BCI-related publications come from the field of psychology.

---

26   Kögel, Johanna, Jennifer R. Schmid, Ralf J. Jox, and Orsolya Friedrich. "Using Brain-Computer Interfaces: A Scoping Review of Studies Employing Social Research Methods." *BMC Medical Ethics* 20, no. 18 (2019). https://doi.org/10.1186/s12910-019-0354-1.

The risks of dependency are real. As one 2025 research paper warned, users "might seek more frequent and intense stimuli to sustain feelings of pleasure, potentially resulting in psychological and behavioral problems such as anxiety, depression, and social withdrawal." The paper cautioned that users "may find ordinary life experiences less fulfilling compared to the heightened sensations provided by BCIs." If this sounds familiar, it echoes the patterns we've already seen with social media addiction—but with far deeper neural integration.

There are also existential-level concerns. A 2021 paper in the *Journal of Futures Studies* identified BCIs as a new risk factor for global totalitarianism, noting that they "allow for an unparalleled expansion of surveillance" and could enable states to "surveil even the mental contents of their subjects" or use brain stimulation to punish dissenting thoughts. The RAND Corporation has similarly warned that "BCI could be misused for totalitarian control of people" and that "people may be psychologically harmed if 'superhuman' capabilities are revoked."

The BCI is, in many ways, the metaverse on steroids—a deep, invasive fusion with the wonderfully complex yet fragile human brain. While it holds the potential to unlock new frontiers of philanthropy, enhancing how we give, collaborate, and solve problems, it also demands careful ethical stewardship. The challenge will not be just in developing this technology but in ensuring it strengthens, rather than erodes, human agency, emotional intelligence, and our ability to find stillness in an always-connected world.

For philanthropy, the questions are urgent: How do we ensure BCIs expand human connection rather than replace it? How do we prevent a "neuro-enhancement divide" where only the wealthy can access cognitive augmentation for charitable leadership? And how do we preserve the contemplative, spiritual dimensions of giving when thoughts themselves become transactional? These are questions we must answer now—before the technology answers them for us.

Emerging Norms and Ethical Frameworks

As artificial intelligence increasingly shapes fundraising practices, the nonprofit sector faces the critical task of integrating AI responsibly to maintain donor trust and uphold ethical standards. Emerging norms and frameworks are guiding organizations on how to navigate data privacy, transparency, bias mitigation, and donor consent, ensuring that AI serves to enhance rather than undermine philanthropic goals.

Responsible AI Use in Fundraising

Key considerations for responsible AI use in fundraising include data privacy and security, which involves protecting donor information and ensuring compliance with relevant privacy regulations. Transparency is also crucial, requiring organizations to be open about how AI tools are deployed in fundraising efforts. Additionally, bias mitigation must be actively addressed by identifying and correcting any biases in AI algorithms that could negatively impact donor engagement strategies. Finally, donor consent is essential, meaning explicit permission should be obtained from donors before their data is used in AI-driven campaigns.

The Association of Professional Researchers for Advancement (APRA) has developed an "Ethics in AI for Fundraising Toolkit" that addresses these issues, providing guidance on compliance and ethical considerations in prospect development. The Fundraising.AI initiative has launched a framework aimed at maximizing the benefits of AI in fundraising while safeguarding public trust. This framework provides guidelines for ethical AI use, emphasizing the need for transparency, accountability, and donor-centric practices. Safeguarding the public trust and the charity brand's positive identity will become tantamount to a charity's survival.

Donor Perceptions and Trust

Understanding donor attitudes toward AI is crucial for its successful integration into fundraising strategies. Studies have shown mixed reactions. A study reported by The Chronicle of Philanthropy found that nearly a third of respondents would be less likely to donate to charities that used AI, highlighting concerns about privacy and the impersonal nature of AI-driven interactions. Conversely, research from Fidelity Charitable indicates that when AI is used to enhance efficiency and impact and when organizations are transparent about its use, donors may respond positively.

The Balance Between Human and Machine

The question isn't whether robots will become part of charitable work—they already are. The question is how we design that integration for human flourishing. We must preserve meaningful work. We must design for human control. We must measure human outcomes—dignity, connection, purpose—not just efficiency.

The writer Charlie Warzel captured a widespread anxiety when he observed that "workplace AI feels like the purest distillation of a corrosive ideology that demands frictionless productivity from workers: The easier our labor becomes, the more of it we can do, and the more of it we'll be expected to do." This fear is legitimate. If we design charitable robotics as efficiency machines—squeezing more output from fewer humans—we will have built precisely the wrong thing.

But there's another path. What if the robot isn't there to replace human effort, but to extend human reach? What if the teleoperator in Boston isn't a worker being automated out of meaning, but a community steward projecting care across continents? What if the humanoid assistant in a home health setting isn't displacing a caregiver, but giving that caregiver more time for the human connection that no machine can provide?

The Opportunities Before Us

The applications are vast and varied:

International aid: Teleoperations can deliver medical supplies to conflict zones, construct emergency shelters after earthquakes, and maintain infrastructure in regions too dangerous for sustained human presence. Each mission builds the collective intelligence of the system—sensorimotor stigmergy in action—while keeping human judgment at the center of every critical decision.

Nature restoration: Robotic systems guided by remote operators can plant trees at scale, remove invasive species, monitor wildlife, and restore ecosystems in terrain too rugged or remote for consistent human access. The operators aren't replaced; they're amplified—one skilled conservationist guiding a fleet of restoration robots across a watershed. Moreover, this will work in cities too, as helpmates in urban gardens or tree wardens in city parks.

Home health: Humanoid assistants can handle the physical labor of caregiving—lifting, transferring, fetching—while freeing human caregivers to provide what matters most: presence, conversation, emotional support. The robot handles the task; the human provides the care. The humanoid is relatable, but one of its chief function is to foster real human relationship – including through communications with friends and family.

Community resilience: Local operators trained to pilot humanitarian robots become skilled workers in their own communities, building capacity that remains long after the crisis passes. Charity autonomy means the tools stay in community hands, and the learning stays at the edge—not extracted to distant servers. Digging deeper, one must ask how these community robots will help us re-build civil society, not hasten it's dissolution.

In each case, the design principle is the same: robots extend human agency rather than replace human purpose.

A New Measure of Success

The coming wave of robotics, teleoperations, and humanoids will surely change how we think about personal contributions. The workplace will be different; we will not only work side by side with AI but also interact with humanoid assistants. How might these assistants augment our good work versus rob us of what is meaningful?

The answer depends on what we optimize for. If we measure success only in packages delivered, hours saved, or costs reduced, we will build systems that treat human involvement as friction to be eliminated. But if we measure success in human terms—dignity preserved, connections deepened, purpose fulfilled, communities empowered—we will build something different. We will build systems where the machine handles what machines do well, and humans remain at the center of what only humans can do.

This is the promise of neuromorphic philanthropy: a charitable ecosystem that mimics not the factory, but the brain. Intelligence distributed to the edge. Learning that happens in the field, not just the data center. Thresholds that activate human oversight at the moments it matters most. And a collective memory—cairns stacked by countless operators and systems—that guides future action without centralizing control.

Humanoids, robotics, and teleoperations will continue to redefine philanthropy. From delivering critical resources to fostering sustainable development, these technologies not only solve immediate problems but also pave the way for long-term community empowerment. By placing the tools in the hands of community stewards—including through charity autonomy—philanthropy can leverage unprecedented potential to enable individuals to craft their own solutions.

The stones are there to be stacked. The trail is ours to build. The only question is whether we design these systems to serve human flourishing, or allow them to be designed for us. I know which path I'm choosing.

Ask Yourself

1. What tasks in your daily life drain your time and energy without enriching your sense of purpose? Could robotic assistance free you to contribute more fully to the causes and relationships that matter most?

2. If a teleoperator in your city could guide a humanitarian robot across the globe, would you consider that meaningful work? What makes work purposeful—the physical presence, the skill applied, or the outcome achieved?

For Charities

1. What roles in your organization are fundamentally about human presence—the work that creates dignity, belonging, and trust? These are the jobs to protect fiercely from automation, not because robots couldn't do the tasks, but because the *relationship* is the point.

2. What physical labor, repetitive logistics, or hazardous tasks currently consume your team's time and energy? Could teleops or humanoid assistance free your people to do more of what only humans can do—listening, counseling, celebrating, grieving alongside those you serve?

2055 - A WORLD UNITED

2055: A World United

Augustina "Gus" Jones arrived just in time for the ceremony. Her electric vehicle hovercraft glided effortlessly above Singapore's tree-lined streets, descending near the main gates of the city's vibrant Botanic Gardens. The Gardens were known for their beauty—especially the internationally famous orchids. Gus thought about Singapore's sophistication both as an Outdoor City and as a nation founded in 1965 with a cohesive vision and unrivalled citizen energy.

While AI agents had entered the Web 3.0 conversation back in 2024, the agents of 2050 were no longer digital abstractions. Especially in Singapore, where universities like Nanyang had taken the lead in their invention and deployment. AI was now embodied—physical, present, and purposefully quiet. Unlike the loud, clunky humanoids Gus had seen in childhood films, these agents moved with a kind of serene service ethic. As one humanoid held the Botanic Garden's gate open, another gracefully handed Gus an augmented reality program book.

Today, Gus was here to receive the Global Innovation in Action Award (GIAA)—one of the world's most coveted honors, given to individuals whose ideas had not only reshaped systems but improved life for humanity and the planet. In an age overflowing with generative content, it wasn't ideas alone that mattered—it was execution, impact, and integrity.

The five-hour flight from Boston to Singapore was now one of the longest flights one could take in the world. In earlier decades, travelers would have flown between those cities by plane, a journey that would have taken nineteen hours from New York. Transportation, housing, food, entertainment, and life in general had changed exponentially in just a few short decades. There was much to celebrate, particularly around humanity's contributions toward the common good.

The GIAA itself was fittingly symbolic. Crafted from coral and driftwood, its design appeared abstract from afar. But up close, one could see its Sashimono-inspired joints—an ancient Japanese wood-joining technique—interlocking like puzzle pieces. These joints symbolized the essence of GIAA: different elements working in harmony, leveraging every tool available to create the greatest good.

Gus had learned that lesson early. Working alone, she'd had moments of brilliance—especially with the power of agentic philanthropy by her side. But working in isolation, even with high-performing tools, limited the reach and resonance of her impact. It was GiveIQ that taught her the difference: that true transformation came not from isolated excellence, but from shared purpose and coordinated action.

Now, standing at the podium with seven other volunteers from around the globe, Gus accepted the award as a team. The proclamation read:

66 ⎯⎯⎯⎯⎯⎯⎯⎯

"When humankind thought the cause was lost, you stepped up and showed us how to respond. That transformational intelligence knows no boundaries. Rather, it is powered by purpose—and the genius of what we can do together."

⎯⎯⎯⎯⎯⎯ **99**

That, Gus knew, was the heart of GiveIQ.

The Age of Purpose Awaits

The tools we have today are merely the beginning. This is the age of early artificial intelligence, the initial forms of Web 3.0, the first indications of quantum computing, and an expansive frontier of tools we have yet to envision. Biological breakthroughs will extend human life. Neuromorphic systems will process information like brains.[1] Humanoid robots will walk among us. The pace of change will not slow down.

In the face of all this, what remains constant?

You do. Your values do. Your capacity to choose purpose over drift, contribution over consumption, connection over isolation—that doesn't get automated. That's yours.

The question is no longer whether we can make a significant difference. The question is whether we will.

The Paradox of Purpose

Here's what I've learned in thirty years of nonprofit work, from building houses with Habitat for Humanity to leading organizations across conservation, workforce development, and now Scouting: the surest path to your own wellbeing runs directly through service to others. Not around it. Through it.

We live in an age of unprecedented anxiety. Loneliness is epidemic. Meaning feels scarce. And yet the research is unambiguous: people who give—their time, their talent, their treasure, their trust—report higher life satisfaction, stronger social bonds, better physical health, and greater resilience in the face of adversity.[2] Purpose isn't a luxury. It's a lifeline.

The wave of AI and technological change will force each of us to choose: to contribute, or to consume. To be swept along, or to plant our feet and build something that matters. Now is the time to set personal and societal guardrails. Not to resist technology, but to align it with what it means to live with purpose, to be fully human, and to give.

The Call to Purpose

Technology alone will not save us. It will not teach us how to care, how to sacrifice, or how to build a just world. That responsibility remains ours.

In this age of accelerating intelligence, we must answer a deeper question: not "What do I do?" but "What impact can I have?" Your Purpose Path keeps you grounded in what's important when everything else is shifting.

This shift from occupation to intention, from task to purpose, is the cornerstone of the world we must now build.

We are entering an era where philanthropy is no longer a privilege. It is a practice. It will be woven into the apps we use, the businesses we support, the agents we train, and the stories we tell. The 4 T's—time, talent, treasure, and trust—aren't just categories of giving. They're the foundation of a life oriented toward meaning in a world that will increasingly tempt us toward distraction.

EIC: Your True North

The EIC framework—Empower, Innovate, Collaborate—isn't just for charities. It's a compass for anyone navigating technological disruption with their humanity intact. And at its core, EIC is about three forms of agency that no technology can replicate.

Empower is the agency to act. Take your GiveIQ assessment. Know where you stand. Then move. Not someday. Now. The agency to act means building your own capacity and helping others build theirs. Not doing for them, but equipping them to act with their own power, amplified by agentic philanthropy.

Innovate is the agency to create. Don't give the way your grandparents did. Bring your full creativity to the problems you care about. Explore AI tools that match you with causes aligned to your values. Ask what problems you're uniquely positioned to solve. The agency to create means refusing to accept that the old answers are good enough.

Collaborate is the agency to connect. The Global Charity Commons grows stronger when diverse perspectives converge. Partner across sectors. Build trust before you build programs. Add a stone to the cairn for those who follow. The agency to connect means your impact doesn't end with your contribution—it multiplies through others.

These three forms of agency hold whether you're an individual discovering your Purpose Path or an organization charting its future. They scale from the personal to the planetary. And they're what remains distinctly human when everything else is shifting.

For Charities: The Stakes Are Higher

If you lead a nonprofit, the coming decade will test everything. AI will reshape how donors give, how programs deliver, and how impact gets measured. Robotics and teleoperations will change what's physically possible. New funding models will emerge. Old ones will fade.

Your job is to navigate this while keeping your mission—and your humanity—intact.

The EIC framework offers a way forward:

Empower your community. Charity Autonomy means putting tools and decision-making power in the hands of those you serve. Not charity as rescue, but charity as capacity-building. What would it look like for your beneficiaries to become co-creators of solutions rather than recipients of services? And as AI, Web 3.0, quantum, and other tools emerge, wisely use them to encourage a networked collective intelligence and combinatorial societal impact.

Innovate relentlessly. Agentic Islands—bounded 90-day experiments—let you test new approaches without betting the organization. What's one pilot you could launch this quarter? One AI tool you could trial? One partnership you could explore? Provide the white space for others to participate. Foster access to a diversity of ideas and opinions—with the vision of a democratization of philanthropy.

Collaborate for the commons. Every transparent impact report, every honest post-mortem on what didn't work, every shared dataset and open-source tool—these are cairns on the trail. They benefit not just your organization but every charity that follows. How are you contributing to the Global Charity Commons, not just drawing from it?

The organizations that thrive will be those that treat technological change not as a threat to be survived, but as an opportunity to deepen their human mission. The 4 T's still matter—perhaps more than ever. Time given by staff and volunteers. Talent deployed strategically. Treasure stewarded transparently. Trust as foundational to all, in an increasingly artificial world. These are the currencies of the age of AI.

The Recklessness of Faith

As William Sloane Coffin said: "I love the recklessness of faith. First you leap, and then you grow wings."[3]

That's what GiveIQ asks. Not perfection. Not certainty. Just the willingness to leap—to commit to an Agentic Island, to walk your Purpose Path, to stack a stone on the trail even when you can't see where it leads. The wings will come.

The Cairns You Leave

I've walked a lot of trails in my life. In the White Mountains, along the Appalachian ridges, through conservation lands I helped protect. What I've learned is this: the cairns you build today guide hikers you'll never meet. That's the quiet power of Purpose Paths—individual actions that accumulate into collective navigation.

The Global Charity Commons isn't an abstraction. It's every trail marker left by every person who chose service above self. Every transparent impact report. Every honest conversation about what didn't work. Every partnership built on trust rather than transaction. Every AI tool designed to empower rather than replace human judgment. Every hour of teleoperation training data that helps the next humanitarian mission succeed.

Some cairns are physical—community centers, conservation land, educational programs built to outlast their founders. Some are digital—platforms, data, shared knowledge that compounds across the commons. All require the same foundation: human experience, honestly assessed and generously shared.

As AI capabilities accelerate, the temptation will be to optimize the Global Charity Commons for efficiency alone. To remove the slower human elements. To let algorithms decide which trails matter.

Resist this. The commons strengthens when human agency leads and agentic philanthropy supports—never the reverse. When qualia informs data. When lived experience guides machine learning. When the 4 T's anchor us to what technology cannot provide: presence, sacrifice, relationship, trust.

Your Purpose Path is yours to walk. But the cairns you leave belong to everyone who follows. Build them strong. Build them honest. Build them to last.

GiveIQ: A Movement

GiveIQ invites you to act from wherever you are—childhood health, climate, literacy, housing, workforce development, whatever stirs your conscience. Your time, talent, treasure, and trust are urgently needed. This is the time of Charity Autonomy. A time not to wait for permission to do good, but to act with your own human agency now.

Those who have been given much have a choice: build the cairns that guide others forward, or leave no mark at all. The trail is being laid now. The stone you place today becomes the path someone else follows tomorrow.

This is the promise of GiveIQ: that the power to do good belongs to everyone, and that the future of giving is not only about scale but also about meaning.

The agency to act. The agency to create. The agency to connect.

These are yours. They don't get automated. They don't get outsourced. They're what you bring to the trail.

That's GiveIQ. That's the path forward.

Ask Yourself

❑ What is my GiveIQ?

And if you lead a team or organization:

❑ What is our GiveIQ?

Notes

1. Neuromorphic computing mimics the structure and function of the human brain, with distributed processing across small computing elements similar to neurons. See:

IBM Research, "What is Neuromorphic Computing?" 2024, https://www.ibm.com/think/topics/neuromorphic-computing; Carver Mead's foundational work at Caltech established the field in the 1980s.

2. Research consistently links prosocial behavior to wellbeing. See: Post, Stephen G., "Altruism, Happiness, and Health: It's Good to Be Good," International Journal of Behavioral Medicine 12, no. 2 (2005): 66-77; Aknin, Lara B., et al., "Prosocial Spending and Well-Being: Cross-Cultural Evidence for a Psychological Universal," Journal of Personality and Social Psychology 104, no. 4 (2013): 635-652.

3. William Sloane Coffin (1924-2006) was an American Christian clergyman and peace activist who served as chaplain at Yale University and senior minister at Riverside Church in New York City. This quote reflects his emphasis on faith as action rather than certainty.

Your Next 90 Days

Don't try to change everything. Start with one cause, one contribution, one measurable action.

Take your GiveIQ assessment this week. See where your agency is flowing and where it's blocked. The matrix won't judge you. It will show you where to begin.

Then choose one Agentic Island. Ninety days. Bounded enough to complete. Significant enough to matter. Pick the area where you've been holding back, the agency you haven't fully exercised, the contribution you've been meaning to make but haven't.

Find one partner. Not because you need help, but because collaboration compounds impact. The agency to connect means your work doesn't end with you.

Document what you do. Honestly. The cairn you leave doesn't have to be impressive. It has to be true.

In three months, take the assessment again. See what moved. Adjust. Repeat.

One cause. One project. One partner. Ninety days.

The wings will come.

APPENDIX:
YOUR 90-DAY
PURPOSE PATH PLAN

Tools for Turning Insight into Action

Your GiveIQ Score shows where you have been active. This plan helps you decide where to go next.

Ninety days is the right unit of time. It is long enough to change behavior and short enough to maintain focus. The goal of this plan is not to 'do more,' but to do deliberately—in your personal life and, if you choose, in your professional one.

Use this section at the end of the book, then return to it every quarter.

How to Use This Plan

- ❑ Review your current GiveIQ Score and EIC × 4T matrix
- ❑ Identify no more than three priority cells for the next 90 days
- ❑ Define specific, evidence-producing actions
- ❑ Consider how your Purpose Path might extend into your workplace
- ❑ Revisit briefly at Day 30 and Day 60
- ❑ Re-score fully at Day 90

This is not a to-do list. It is a capacity-building plan.

The EIC × 4T Planning Grid

A Visual Tool for Your 90-Day Focus

Score each cell 1-10 based on your activity over the past 90 days. Circle your priorities for the coming quarter.

Note: Self-assessment yields a maximum of 120 points. Verified contributions (peer-validated, organization-confirmed, or impact-documented) can multiply your score up to 360 in the full GiveIQ system.

	TIME *Your hours and attention*	**TALENT** *Your skills and expertise*	**TREASURE** *Your financial resources*	**TRUST** *Your relationships*
EMPOWER *Building your own capacity*	Learn, prepare, study Score: ___ ☐ Priority	Develop, practice, train Score: ___ ☐ Priority	Invest in your growth Score: ___ ☐ Priority	Build capacity-expanding relationships Score: ___ ☐ Priority
INNOVATE *Applying creativity and expertise*	Focused problem-solving Score: ___ ☐ Priority	Apply expertise in new ways Score: ___ ☐ Priority	Fund experim-entation Score: ___ ☐ Priority	Engage those with fresh perspectives Score: ___ ☐ Priority
COLLABORATE *Creating shared impact*	Show up with others Score: ___ ☐ Priority	Contribute skills to collective effort Score: ___ ☐ Priority	Pool resources for greater effect Score: ___ ☐ Priority	Connect, introduce, convene Score: ___ ☐ Priority
SUBTOTAL	___/30	___/30	___/30	___/30

Pillar Totals: Empower ___/40 | Innovate ___/40 | Collaborate ___/40

Self-Assessment Score: ___/120

With Verification Multipliers, Maximum Possible: ___/360

Reading Your Patterns

If your scores cluster in one row: You have a strong orientation toward that pillar. Consider whether the others deserve attention.

If your scores cluster in one column: You tend to contribute through that resource. Consider diversifying.

If scores are even across the grid: You may be spreading thin. The Optimizer pathway might serve you.

If one cell dominates: You've found your sweet spot—or your comfort zone. Is it time to stretch?

Where You Are

The EIC × 4T Self-Assessment

Your GiveIQ Score is built on the intersection of three principles (Empower, Innovate, Collaborate) and four resources (Time, Talent, Treasure, Trust). This creates twelve cells, each representing a distinct way you contribute.

Using the visual grid on the previous page, take fifteen minutes to honestly assess your activity over the past 90 days in each cell.

Scoring Guide

- ❑ 1-3: Little or no meaningful activity
- ❑ 4-6: Sporadic or modest engagement
- ❑ 7-9: Consistent, sustained contribution
- ❑ 10: Exceptional contribution with documented impact

Verification Multipliers

Self-assessment is the starting point, but verified activity carries more weight:

Self-reported: Score × 1.0 | Peer-validated: Score × 1.25

Organization-confirmed: Score × 1.5 | Impact-documented: Score × 2.0

Maximum possible score with full verification: 360 points

Your Level

SCORE	LEVEL	WHAT IT MEANS
0-60	**Awakening**	Beginning your philanthropic journey
61-120	**Emerging**	Engaged but still finding your path
121-180	**Active**	Making real contributions across multiple dimensions
181-240	**Engaged**	Consistent, meaningful contributor
241-300	**Amplified**	High capacity with documented impact
301-360	**Transformative**	Philanthropic leader multiplying impact through others

Pattern Analysis

Strongest pillar: _____ (your natural approach)

Underutilized pillar: _____ (your growth opportunity)

Strongest T: _____

Underutilized T: _____

Insight lives in imbalance. Growth comes from intention.

Where to Begin

Finding Your Starting Point

Based on your score, one of these pathways may resonate. They are not prescriptions—they are starting points.

The Activator (Scores 0-120)

You are early in your journey, or returning after time away. Your 90-day focus: depth over breadth. Choose one cause, one organization, one bounded project. Resist the temptation to spread thin. The goal is not to catch up but to begin with intention.

The Optimizer (Scores 121-240)

You are engaged but perhaps scattered. Your 90-day focus: consolidation. Audit where your energy has gone. Ask which activities produced genuine impact and which merely kept you busy. Decline commitments that dilute your focus. Go deeper with fewer.

The Multiplier (Scores 241-360)

You are operating at high capacity. Your 90-day focus: extending your reach through others. Identify people in your life with untapped philanthropic potential. Invite them in. Walk alongside them. The measure of this season is not your own output but whether others find their footing.

What to Do

From Intention to Action

Your Intention

In one sentence, describe what you want this next quarter to represent:

Over the next 90 days, I intend to:

_____

_____

_____

Priority Cells

You do not need to work all twelve cells at once. Select one to three areas where growth will matter most. Leave the rest for another season.

PRIORITY	PILLAR	T	WHY THIS CELL MATTERS NOW
1	☐ Empower ☐ Innovate ☐ Collaborate	☐ Time ☐ Talent ☐ Treasure ☐ Trust	
2	☐ Empower ☐ Innovate ☐ Collaborate	☐ Time ☐ Talent ☐ Treasure ☐ Trust	
3	☐ Empower ☐ Innovate ☐ Collaborate	☐ Time ☐ Talent ☐ Treasure ☐ Trust	

Action Planning

For each priority cell, define one concrete action and the evidence it will produce.

Priority 1:

Planned action: _____

Target evidence: _____

Priority 2:

Planned action: _____

Target evidence: _____

Priority 3:

Planned action: _____

Target evidence: _____

Charity Autonomy

Agentic Philanthropy

Agentic Islands

Global Charity Commons

Your Agentic Island

An Agentic Island is a bounded project with a clear beginning, end, and deliverable. It is how you move from general goodwill to specific contribution.

Project Name: _____

Organization: _____

Duration: _____ days (typically 60-120)

Specific Deliverable:

(Not 'help with fundraising' but 'create donor cultivation plan for 50 lapsed major donors')

_____

Weekly Commitment: _____ hours

Primary Contact: _____

Success Criteria:

(How will you and the organization know this succeeded?)

_____

Handoff Plan:

(How will your work continue after you step back?)

_____

Start Date: _____   End Date: _____

Your Purpose at Work

Extending Your Path Into Professional Life

Your Purpose Path does not pause when you arrive at the office.

For most of us, work occupies a significant portion of our waking lives and shapes how we see ourselves in the world. The question is not whether your professional life intersects with your philanthropic identity—it is whether that intersection happens by accident or by design.

This section invites you to consider how your 90-day focus might extend into your workplace, and how you might gently influence the culture around you.

Three Modes of Integration

Personal Practice

This is the most accessible layer: how you individually engage through the structures your employer already offers. Many organizations provide volunteer time off, matching gift programs, or skills-based service opportunities. These benefits often go underutilized—not from lack of interest, but from lack of intention.

Consider: What exists at your workplace that you have not fully explored? What would it mean to treat these programs not as perks but as extensions of your Purpose Path?

Your 90-day action:

_____

_____

_____

_____

Team Influence

You likely work alongside people who care about making a difference but have not yet found their entry point. The Multiplier pathway applies here, too. An invitation from a trusted colleague often carries more weight than a corporate announcement.

This does not require organizing elaborate initiatives. It might mean sharing why a particular cause matters to you, inviting a teammate to join you for a volunteer morning, or simply asking others what they care about and listening to the answer.

Your 90-day action:

_____

_____

Organizational Advocacy

This is the longer work. Most corporate social responsibility programs, however well-intentioned, remain rooted in a transactional model: dollars donated, hours logged, reports published. The principles at the heart of GiveIQ—Empower, Innovate, Collaborate—suggest a different possibility.

You do not need positional authority to influence this conversation. You need thoughtful questions and patience. In meetings where community engagement arises, you might wonder aloud: Are employees empowered to shape our giving, or simply invited to participate? Do we approach nonprofit partners as collaborators or as recipients? Is there room for experimentation in how we engage, not just what we fund?

These questions rarely produce immediate change. But they plant seeds. Over time, they can shift what an organization believes is possible.

Your 90-day action:

_____

_____

Where Does Your Organization Currently Sit?

This simple framework may help you locate your workplace's current approach—and imagine where it might go.

	TRADITIONAL CSR	EMERGING PRACTICE	GIVEIQ-ALIGNED
GIVING	Corporate foundation or leadership decides	Employee matching available	Employee-directed, evidence-informed
VOLUNTEERING	Annual day of service	Flexible PTO for volunteering	Skills-based, project-bounded engagement
MEASUREMENT	Dollars donated, hours logged	Participation rates tracked	Impact evidence, capacity built
EMPLOYEE ROLE	Participant	Chooser	Agent

Most organizations live somewhere between Traditional and Emerging. The opportunity is not to revolutionize your company in a single quarter. It is to move one conversation, one practice, one assumption gently forward.

A Workplace Agentic Island

The Agentic Island concept applies in professional settings, too. Rather than advocating abstractly for better corporate engagement, you might design a bounded project that demonstrates what is possible—one that advances both organizational goals and your Purpose Path.

Project concept: _____

Organizational benefit: _____

Community benefit: _____

Duration: _____ weeks

Colleagues to involve: _____

What success looks like:

_____

_____

How this advances your Purpose Path:

_____

_____

How to Learn

Staying Honest and Learning Forward

Staying Honest With Yourself

Before you begin, review these common patterns. They are not failures—they are tendencies we all share. Awareness is the first defense.

- ❑ **The Passion Trap** — Spreading energy across too many causes because each one matters.
  The remedy: Choose one. Just one. For 90 days.

- ❑ **The Savior Complex** — Arriving with solutions before understanding the problem.
  The remedy: Spend the first weeks listening.

- ❑ **The Disappearing Volunteer** — Fading away when life gets busy, without communicating.
  The remedy: If you need to pause, say so directly. Organizations can plan around honesty.

- ❑ **The Credential Collector** — Joining boards and committees without clear contribution.
  The remedy: Before saying yes, ask: What will I specifically accomplish?

- ❑ **The Lone Wolf** — Preferring independent work over the inefficiency of collaboration.
  The remedy: Invest in relationships, even when it slows you down.

Mid-Cycle Check-In (Day 30 and Day 60)

Take ten minutes at each milestone to reflect:

What is working?

_____

_____

_____

What needs adjustment?

_____

_____

_____

What has surprised you?

_____

_____

90-Day Reflection

At the end of the quarter, return to your EIC × 4T matrix and score again.

New GiveIQ Score: _____ / 360

Change from last cycle: _____

Which cells showed the most growth?

_____

_____

_____

_____

Which remained unchanged, and why?

_____

_____

_____

_____

What pattern is emerging about how you create impact?

_____

_____

_____

What will you carry into the next 90 days?

_____

_____

_____

Evidence of Your Journey

For each priority cell, how was your contribution verified?

CELL	EVIDENCE TYPE
1	☐ Self-reported ☐ Peer-validated ☐ Organization-confirmed ☐ Impact-documented
2	☐ Self-reported ☐ Peer-validated ☐ Organization-confirmed ☐ Impact-documented
3	☐ Self-reported ☐ Peer-validated ☐ Organization-confirmed ☐ Impact-documented

Evidence turns good intentions into portable credibility. It allows your story to travel beyond your own telling.

Your GiveIQ Score is not about keeping score.

It is about leaving a trail others can follow.

RETURN TO THIS PLAN AT THE START OF EACH QUARTER. OVER TIME, THE PATTERNS WILL REVEAL NOT JUST WHAT YOU DO, BUT WHO YOU ARE BECOMING.

PART TWO: FOR NONPROFITS

Building a GiveIQ Organization

A 90-Day Framework Integrating the 72-Point Assessment™

EIC × 4 T's × 6 Business Units

Empower ✦ *Innovate* ✦ *Collaborate*
E: Agency to Act. I: Agency to Create. C: Agency to Connect.
Multiplied by: Time ✦ *Talent* ✦ *Treasure* • *Trust*

From the GiveIQ Framework™ by John Judge

Introduction: From Assessment to Action

The GiveIQ framework serves both individuals finding their Purpose Path and nonprofit organizations seeking to build cultures where staff, volunteers, board members, and community partners can thrive together. This appendix bridges the comprehensive 72-Point Assessment with practical organizational implementation.

A GiveIQ organization does three things well: it **empowers** everyone in its ecosystem to contribute meaningfully, it **innovates** continuously in how work gets done, and it **collaborates** openly with the broader philanthropic community.

Improving Purpose Paths for Your Community

Just as hikers leave cairns—those carefully stacked stones—to guide those who follow, GiveIQ organizations have a responsibility beyond their own operations. Every system you build, every lesson you learn, every innovation you develop becomes a potential cairn for others walking similar Purpose Paths. When you share these openly through the Global Charity Commons, you're not just improving your organization— you're improving the trail for everyone who comes after.

The most powerful organizations understand this dual mandate: excel in your own mission *and* leave the philanthropic landscape better than you found it. The 72-Point Assessment measures both—your internal effectiveness and your contribution to the commons.

How the Frameworks Connect

The full 72-Point Assessment evaluates organizations across a 3 × 4 × 6 matrix. This organizational framework translates that comprehensive assessment into three operational domains:

EIC PRINCIPLE	DOMAIN	CORE QUESTION
EMPOWER	Culture	Does your organization enable Charity Autonomy for all?
INNOVATE	Infrastructure	Does your infrastructure support experimentation and learning?
COLLABORATE	Program	Does programming embrace the Global Charity Commons?

GiveIQ Ecosystem Readiness Assessment

Before diving into the full 72-point assessment, use this simplified readiness assessment. Score each dimension 1-10 using: Absent (1-2), Emerging (3-4), Developing (5-6), Established (7-8), Leading (9-10).

EMPOWER: Culture Assessment

72-Point Assessment: Points 1-4, 13-16, 25-28, 37-40, 49-52, 61-64

DIMENSION	SCORE	EVIDENCE
Barrier Removal How easily can new contributors begin meaningful work?	___/10	
Clear Pathways Are there defined ways to grow in responsibility?	___/10	
Meaningful Recognition Is contribution acknowledged in ways that matter?	___/10	
Distributed Decision-Making Do those closest to the work have real authority?	___/10	
EMPOWER SUBTOTAL	___/40	

INNOVATE: Infrastructure Assessment

72-Point Assessment: Points 5-8, 17-20, 29-32, 41-44, 53-56, 65-68

DIMENSION	SCORE	EVIDENCE
Agentic Island Portfolio Do you maintain well-designed, bounded projects?	___/10	
Skill-to-Need Matching Are talents matched intelligently to needs?	___/10	

DIMENSION	SCORE	EVIDENCE
AI-Enabled Coordination Does technology amplify rather than bureaucratize?	___/10	
Learning Loops Do systems capture and apply lessons?	___/10	
INNOVATE SUBTOTAL	___/40	

COLLABORATE: Program Assessment

72-Point Assessment: Points 9-12, 21-24, 33-36, 45-48, 57-60, 69-72

DIMENSION	SCORE	EVIDENCE
Common Data Standards Do you use and contribute to shared measurement?	___/10	
Cross-Organization Projects Do you actively partner with peers on shared goals?	___/10	
Transparent Resource Flows Can stakeholders see how resources are used?	___/10	
Collective Learning Do you share successes and failures openly?	___/10	
COLLABORATE SUBTOTAL	___/40	

Total Ecosystem Readiness

Empower: ___/40 + Innovate: ___/40 + Collaborate: ___/40

TOTAL READINESS SCORE: ____/120

SCORE	LEVEL	WHAT IT MEANS & NEXT STEP
0-40	**Foundation**	Focus on full 72-Point for one business unit before expanding.
41-80	**Building**	Ready for 90-Day Plan focusing on weakest EIC dimension.
81-100	**Ready**	Conduct full 72-Point across all business units.
101-120	**Leading**	Ready to add cairns to the Global Charity Commons and guide others.

90-Day Implementation Plan

Transformation happens in quarters, not years. Each phase builds momentum that sustains beyond the initial effort. As you progress, remember: every improvement you document becomes a potential cairn for others on similar Purpose Paths.

Phase 1: AUDIT (Weeks 1-3)

Map Contributors by 4 T's

- ❑ **Time:** Staff, volunteers, board, pro bono professionals
- ❑ **Talent:** Skills available, gaps, development needs
- ❑ **Treasure:** Donors, sponsors, grant-makers, customers
- ❑ **Trust:** Relationships, credibility, social capital, reputation

Total contributors identified: _____

Identify Super-Contributors & Potential Agentic Islands

Who contributes most across multiple T's? What bounded projects could you create? Consider how each project might improve Purpose Paths for your community.

Top 5 contributors and their 72-Point indicators touched:

1. _____ | Indicators: _____

2. _____ | Indicators: _____

3. _____ | Indicators: _____

Potential Agentic Islands:

Island 1: _____ | Target Indicator(s): _____

Island 2: _____ | Target Indicator(s): _____

Island 3: _____ | Target Indicator(s): _____

Phase 2: DESIGN (Weeks 4-6)

Agentic Island Brief Template

ISLAND NAME:	
DURATION:	_____ weeks
SPECIFIC DELIVERABLE:	
4 T'S REQUIRED:	☐ Time ☐ Talent ☐ Treasure ☐ Trust
BUSINESS UNIT:	☐ Gov ☐ HR ☐ Fund ☐ Comm ☐ Prog ☐ Fin
TARGET 72-POINT INDICATOR(S):	
EXPECTED SCORE IMPACT:	Current: ___ Target: ___
CAIRN POTENTIAL:	What can we share with the Commons?

Phase 3: PILOT (Weeks 7-12)

Recruit 3-5 pilot contributors, support with weekly check-ins, then debrief and re-score. Document what you learn—these insights become cairns for others.

Key learnings: _____

72-Point score changes: _____

Adjustments for scale: _____

Cairn to share with Commons: _____

Your Organization's GiveIQ Commitment

A GiveIQ organization operates with a dual mandate: excel in your own mission and leave the philanthropic landscape better than you found it. Each commitment below includes space for identifying how you'll contribute to the Global Charity Commons—the cairns you'll leave for others walking similar Purpose Paths.

EMPOWER

Enable Charity Autonomy. Don't be the bottleneck.

Key 72-Point indicators to improve: _____

Our commitment: _____

Cairn for the Commons: _____

INNOVATE

Maintain Agentic Islands. Use AI to coordinate, not control.

Key 72-Point indicators to improve: _____

Our commitment: _____

Cairn for the Commons: _____

COLLABORATE

Participate in the Global Charity Commons.

Key 72-Point indicators to improve: _____

Our commitment: _____

Cairn for the Commons: _____

IMPACT

100% improvement in five years. Better Purpose Paths for all.

Our impact metric: _____

Our commitment: _____

How we'll improve Purpose Paths beyond our organization:

_____

_____

_____

Signed:

Name: _____ Title: _____

Organization: _____ Date: _____

The distance between intention and impact is action.

The distance between action and transformation is consistency.

The distance between your journey and another's is a cairn.

START TODAY.

GiveIQ 72 Point Assessment available at: www.giveiq.com

ACKNOWLEDGMENTS

I'd like to thank everyone who has bounced AI and technology ideas around with me over the past three years. And I'm especially grateful to those who have supported me and helped me during an incredibly rewarding nonprofit career over the last thirty years. Those mentors in nonprofit organizations and philanthropy who exemplified what doing good work was all about. I have been really blessed by all of the people I have had the chance to work and partner with along the way.

I tried to create an easy-to-use book that helps one become a lifelong philanthropist and realize their Purpose Path and potential. I also wrote this book to help nonprofits have GiveIQ as a cornerstone of their next generation. It's harder than ever to be a nonprofit CEO – including the demands to keep up with technological change. The EIC pillars can keep one grounded in what's important and build from there.

Many of us now use AI without defaulting to it. That distinction matters. With our human agency protected, agentic philanthropy becomes a force multiplier for lives of purpose and impact. The expanding digital world can engross us, or we can choose to step through it and into real, lasting change.

The ideas in this book emerged from thirty years of working alongside volunteers, donors, and nonprofit leaders, from hauling lumber with President Carter in the Philippines to navigating Scout council mergers in New England. Every framework, every principle, every argument is rooted in that lived experience.

In the writing process, I used AI as a collaborative tool: for research, for editing, for stress-testing ideas, for turning my hand drawn illustrations into digital art, and for helping transform a sprawling draft into a more focused manuscript. It's fitting, perhaps, that a book about AI amplifying human purpose was itself written with AI amplifying mine.

BIBLIOGRAPHY

Abnormal Security. "Nonprofits Face Surge in Cyber-Attacks as Email Threats Rise 35%." *Infosecurity Magazine*, March 5, 2025. https://www.infosecurity-magazine.com/news/nonprofits-email-threats-rise-35/.

Aknin, Lara B., Christopher P. Barrington-Leigh, Elizabeth W. Dunn, John F. Helliwell, Justine Burns, Robert Biswas-Diener, Imelda Kemeza, Paul Nyende, Claire E. Ashton-James, and Michael I. Norton. "Prosocial Spending and Well-Being: Cross-Cultural Evidence for a Psychological Universal." *Journal of Personality and Social Psychology* 104, no. 4 (2013): 635–652.

Amazon Web Services. "AWS IoT TwinMaker: Digital Twin Technology." Seattle: Amazon Web Services, 2024. https://aws.amazon.com/iot-twinmaker.

Apple Inc. U.S. Patent Application for "Earbud with Biosensor." Cupertino, CA: Apple Inc., 2024.

Association of Professional Researchers for Advancement. "Ethics in AI for Fundraising Toolkit." APRA, 2024. https://www.aprahome.org.

BDO. "The Crucial Role of Cybersecurity for Nonprofit Organizations in 2025." BDO Insights, February 19, 2025. https://www.bdo.com/insights/industries/nonprofit-education/.

BoardEffect. "Nonprofits and Cyberattacks: Key Stats That Boards Need to Know." BoardEffect Blog, 2024. https://www.boardeffect.com/blog/nonprofits-cyberattacks-key-stats/.

Boston Dynamics. "Atlas and Spot: Advanced Robotics for Real-World Applications." Waltham, MA: Boston Dynamics, 2024. https://www.bostondynamics.com.

Brightcove. "The Science of Social Video: Turning Viewers into Advocates." Boston: Brightcove, 2018.

Buterin, Vitalik, Zoë Hitzig, and E. Glen Weyl. "A Flexible Design for Funding Public Goods." *Management Science* 65, no. 11 (2019): 5171–5187.

Carter Center. "Guinea Worm Eradication Program." Atlanta: The Carter Center, 2024. https://www.cartercenter.org/health/guinea_worm/index.html.

CCS Fundraising. "2025 Philanthropic Landscape: Trends and Outlook." New York: CCS Fundraising, 2025.

Child, Julia, with Simone Beck and Louisette Bertholle. *Mastering the Art of French Cooking*. New York: Alfred A. Knopf, 1961.

CLA Connect. "Cybersecurity for Nonprofits in the Age of AI-Based Attacks." CLA Nonprofit Blog, 2025. https://www.claconnect.com/en/resources/blogs/nonprofits/.

Coffin, William Sloane. *Credo*. Louisville, KY: Westminster John Knox Press, 2004.

Community IT Innovators. "Nonprofit Cybersecurity Stats: 10 Numbers to Know." August 16, 2024. https://communityit.com/nonprofit-cybersecurity-stats-10-numbers-to-know/.

Diamandis, Peter H., and Steven Kotler. *Abundance: The Future Is Better Than You Think*. New York: Free Press, 2012.

Dixon, Chris. *Read Write Own: Building the Next Era of the Internet*. New York: Random House, 2024.

DonorsChoose. "How DonorsChoose Works: Connecting Donors with Classrooms." New York: DonorsChoose, 2024. https://www.donorschoose.org.

DreamBox Learning. "Intelligent Adaptive Learning for K-8 Math." Bellevue, WA: DreamBox Learning, 2024. https://www.dreambox.com.

European Parliament. "Brain-Computer Interfaces: Current Applications and Future Prospects." European Parliamentary Research Service, 2024.

Federal Trade Commission. "Gift Card Scams Cost Consumers $217 Million in 2023." FTC Consumer Protection Data Spotlight, March 2024. https://www.ftc.gov/data.

Fidelity Charitable. "Donor Perceptions of AI in Fundraising." Boston: Fidelity Charitable, 2024.

Fundraising.AI. "Framework for Ethical AI Use in Fundraising." 2024. https://www.fundraising.ai.

Galen, Doug, Nikki Brand, Lyndsey Boucherle, Rob Davis, Natalie Do, Ben El-Baz, Isadora Kimura, Kate Wharton, and Jay Lee. "Blockchain for Social Impact:

Moving Beyond the Hype." Stanford Graduate School of Business Center for Social Innovation, 2018. https://www.gsb.stanford.edu/faculty-research/publications/blockchain-social-impact.

Gartner. "Agentic AI Will Autonomously Resolve 80% of Common Customer Service Issues by 2029." Press release, 2024.

Gilbert, Frederic, et al. "Embodiment and Estrangement: Results from a First-in-Human 'Intelligent BCI' Trial." *Science and Engineering Ethics* 25, no. 1 (2019): 83–96.

Gitcoin. "Gitcoin Grants: Funding Open Source and Public Goods." 2024. https://www.gitcoin.co/grants.

Google. "Google for Nonprofits Will Expand to 100+ New Countries and Launch 10+ New No-Cost AI Features." The Keyword, June 11, 2025. https://blog.google/outreach-initiatives/google-org/.

Google. "NotebookLM Adds Deep Research and Support for More Source Types." The Keyword, November 15, 2025. https://blog.google/technology/google-labs/notebooklm-deep-research-file-types/.

Grand View Research. "Agentic AI Tools Market Size Report, 2025–2030." San Francisco: Grand View Research, 2025.

Guterres, António. "Remarks at the UN Climate Change Conference (COP27)." Sharm El-Sheikh, Egypt, November 2022.

Honda Motor Company. "ASIMO: The World's Most Advanced Humanoid Robot." Tokyo: Honda Motor Company, 2024. https://global.honda/innovation/robotics/ASIMO.html.

Ibarz, Julian, Jie Tan, Chelsea Finn, Mrinal Kalakrishnan, Peter Pastor, and Sergey Levine. "How to Train Your Robot with Deep Reinforcement Learning: Lessons We Have Learned." *The International Journal of Robotics Research* 40, no. 4–5 (2021): 698–721.

IBM Research. "What is Neuromorphic Computing?" 2024. https://www.ibm.com/think/topics/neuromorphic-computing.

iNaturalist. "Citizen Science for Biodiversity." California Academy of Sciences and National Geographic Society, 2024. https://www.inaturalist.org.

Intuition Robotics. *ElliQ Impact Report: Combating Loneliness Through AI*. Ramat Gan, Israel: Intuition Robotics, 2023.

Isaacson, Walter. *Steve Jobs*. New York: Simon & Schuster, 2011.

Judge, John. *The Outdoor Citizen: Get Out, Give Back, Get Active*. Guilford, CT: FalconGuides, 2019.

Krauhausen, Imke, et al. "Organic Neuromorphic Electronics for Sensorimotor Integration and Learning in Robotics." *Science Advances* 8, no. 35 (2022).

Kurzweil, Ray. *The Singularity Is Nearer: When We Merge with AI*. New York: Viking, 2024.

Marler, Timothy, et al. "Brain-Computer Interface Technologies: Opportunities and Challenges." Santa Monica, CA: RAND Corporation, 2024.

Maslow, Abraham. "A Theory of Human Motivation." *Psychological Review* 50, no. 4 (1943): 370–396.

McKinsey & Company. "The State of AI in 2024: Gen AI Adoption Spikes and Starts to Generate Value." McKinsey Global Survey, 2024.

Microsoft Canada. "Attention Spans: Consumer Insights." Spring 2015.

Mullin, Emily. "Neuralink's First Human Patient Can Control a Computer With His Brain." *Wired*, March 20, 2024.

Muro, Mark, Robert Maxim, and Jacob Whiton. "Automation and Artificial Intelligence: How Machines Are Affecting People and Places." Washington, DC: Brookings Institution, January 2019. https://www.brookings.edu/research/automation-and-artificial-intelligence-how-machines-affect-people-and-places/.

National Interagency Fire Center. "Wildland Fire Statistics." Boise, ID: NIFC, 2024. https://www.nifc.gov/fire-information/statistics.

Neuralink Corporation. "First Human Patient Update." Press release, March 2024.

Ng, Andrew. "AI Is the New Electricity." Keynote address, Stanford Graduate School of Business, 2017.

NPR. "Jimmy Carter Took on Guinea Worm Disease—and Had Remarkable Success." December 2024.

PATH. "Digital Health Solutions for Resource-Limited Settings." Seattle: PATH, 2024. https://www.path.org/programs/digital-health.

Post, Stephen G. "Altruism, Happiness, and Health: It's Good to Be Good." *International Journal of Behavioral Medicine* 12, no. 2 (2005): 66–77.

Putnam, Robert D. *Bowling Alone: The Collapse and Revival of American Community*. New York: Simon & Schuster, 2000.

Remitly. "Transforming International Money Transfers." Seattle: Remitly, 2024. https://www.remitly.com/us/en/about.

Replit. "Replit vs Cursor: Which AI Coding Platform Fits Your Workflow?" Replit Discover, 2025. https://replit.com/discover/replit-vs-cursor.

Resident.com. "Free AI Video Generator for Nonprofits: Tell Your Story Effectively." July 9, 2025. https://resident.com/resource-guide/.

Scott, MacKenzie. "Seeding by Ceding." Medium, June 15, 2021. https://mackenzie-scott.medium.com.

Sirangelo, Jennifer. "The State of Volunteering in America." Points of Light, 2024.

SoftBank Robotics. "Pepper: The Emotional Robot." Tokyo: SoftBank Robotics, 2024. https://www.softbankrobotics.com/emea/en/pepper.

Spitzer, Manfred. *Digitale Demenz: Wie wir uns und unsere Kinder um den Verstand bringen*. Munich: Droemer, 2012.

Stanford Social Innovation Review. "Five Ways AI Can Deepen Nonprofit Relationships." Spring 2025.

Superframeworks. "10 Best AI Coding Tools 2025: Vibe Coding Tools Compared." November 8, 2025. https://superframeworks.com/blog/best-ai-coding-tools.

Synthesia. "The 13 Best AI Video Generators (I've Actually Tested)." 2025. https://www.synthesia.io/post/best-ai-video-generators.

Taylor Corporation. "Gift Card Draining and the 2024 Gift Card Scams Prevention Act." Taylor Blog, November 12, 2024. https://www.taylor.com/blog/gift-card-draining/.

Technology Association of Grantmakers. "2024 State of AI in Philanthropy Survey." TAG, 2024.

Tesla, Inc. "Optimus: Tesla Humanoid Robot." Austin, TX: Tesla, 2024. https://www.tesla.com/AI.

The Modern Nonprofit. "Nonprofits Are Prime Targets for Cyberattacks—Is Your Organization at Risk?" March 12, 2025. https://themodernnonprofit.com/.

Ushahidi. "Crisis Mapping Platform." Nairobi: Ushahidi, 2024. https://www.ushahidi.com.

Wang, Xiaofeng, and Kevin C. Desouza. "China's New Model of Blockchain-Driven Philanthropy." *Stanford Social Innovation Review* 17, no. 3 (Summer 2019): 26–31. https://ssir.org/articles/entry/chinas_new_model_of_blockchain_driven_philanthropy.

Warzel, Charlie. "The Age of AI Slop." *The Atlantic*, 2024.

Webb, Amy. *The Signals Are Talking: Why Today's Fringe Is Tomorrow's Mainstream.* New York: PublicAffairs, 2016.

Wiggers, Kyle. "A Nonprofit Is Using AI Agents to Raise Money for Charity." *TechCrunch*, April 8, 2025. https://techcrunch.com/2025/04/08/a-nonprofit-is-using-ai-agents-to-raise-money-for-charity/.

Wiggers, Kyle. "Lovable Reaches $100M ARR in Eight Months." *TechCrunch*, 2025.

World Economic Forum. "The Fourth Industrial Revolution: What It Means, How to Respond." Geneva: World Economic Forum, 2016.

Xu, Nancy. Interview on AI agents. MoonHub, 2024.

Zuckerberg, Mark. "Meta AI and the Future of Personalized Assistants." Meta Connect keynote, 2024.

GiveIQ

🌐 **www.GiveIQ.com**

www.ingramcontent.com/pod-product-compliance
Lightning Source LLC
Chambersburg PA
CBHW071556210326
41597CB00019B/3265